AF891948

Atif Mahboob

Modelling and use of SysML behaviour models for achieving dynamic use cases of technical products in different VR-systems

Berichte aus dem
INSTITUT FÜR MASCHINEN- UND
GERÄTEKONSTRUKTION (IMGK)

Herausgegeben von
Univ.-Prof. Dr.-Ing. Ulf Kletzin (Maschinenelemente),
Univ.-Prof. Dr.-Ing. René Theska (Feinwerktechnik) und
Univ.-Prof. Dr.-Ing. Christian Weber (Konstruktionstechnik)
aus dem Institut für Maschinen- und Gerätekonstruktion (IMGK) an der TU Ilmenau.

Band 38

Diese Reihe setzt die „Berichte aus dem Institut für Maschinenelemente und Konstruktion“ fort.

Modelling and use of SysML behaviour models for achieving dynamic use cases of technical products in different VR-systems

Atif Mahboob

Universitätsverlag Ilmenau
2021

Impressum

Bibliografische Information der Deutschen Nationalbibliothek
Die Deutsche Nationalbibliothek verzeichnet diese Publikation in der Deutschen Nationalbibliografie; detaillierte bibliografische Angaben sind im Internet über http://dnb.d-nb.de abrufbar.

Diese Arbeit hat der Fakultät für Maschinenbau der Technischen Universität Ilmenau als Dissertation vorgelegen.

Tag der Einreichung:	7. April 2020
1. Gutachter:	Univ.-Prof. Dr.-Ing. Christian Weber (Technische Universität Ilmenau)
2. Gutachter/-in:	Univ.-Prof. Dr. phil. Heidi Krömker (Technische Universität Ilmenau)
3. Gutachter:	Univ.-Prof. Dr.-Ing. Rainer Stark (Technische Universität Berlin)
Tag der Verteidigung:	12. November 2020

Technische Universität Ilmenau/Universitätsbibliothek
Universitätsverlag Ilmenau
Postfach 10 05 65
98684 Ilmenau
http://www.tu-ilmenau.de/universitaetsverlag

readbox unipress
in der readbox publishing GmbH
Rheinische Str. 171
44147 Dortmund
http://unipress.readbox.net/

ISSN 2191-8082 (Druckausgabe)
ISBN 978-3-86360-234-5 (Druckausgabe)
DOI 10.22032/dbt.47179
URN urn:nbn:de:gbv:ilm1-2020000595

Dedicated to my family

Foreword by the editor

Since the early 1950s, the design of machines and devices, as well as the associated methods and tools, have been the prime focus at the Technische Universität Ilmenau and its predecessor institutions. It was, therefore, an obvious step that the three design-oriented research groups within the Department of Mechanical Engineering – Machine Elements, Precision Engineering, Engineering Design – founded the Institut für Maschinen- und Gerätekonstruktion (IMGK, Institute of Machine Design and Precision Engineering) in May 2008. The IMGK stands in the tradition of a chain of similar predecessor institutes in changing formations over time, mainly due to the changes in the university structure.

The purpose of the institute is to combine the expertise and resources of the participating research groups in order to enhance research and teaching and to provide the public with its scientific results in a joint effort.

An essential instrument for this is the IMGK's book series. It continues a successful publication series from the immediate predecessor institute IMK (Institut für Maschinenelemente und Konstruktion, Institute of Machine Elements and Engineering Design), which was founded in 1991.

The book series primarily includes the doctoral theses conducted at the institute. Besides, it also includes research reports that fit into the thematic framework and are of general interest.

The present volume 38 at hand was written as a doctoral thesis in the Engineering Design Group under the scientific supervision of Professor Christian Weber.

The editors hope that there will be a keen interest in the publication series and would be pleased if it could contribute to a fruitful dialogue in science and practice.

Ilmenau, November 2020

Univ.-Prof. Dr.-Ing. Ulf Kletzin (Machine Elements)

Univ.-Prof. Dr.-Ing. René Theska (Precision Engineering)

Univ.-Prof. Dr.-Ing. Christian Weber (Engineering Design)

Foreword by the author

This thesis is the outcome of the research carried out during my time as a research assistant in the Engineering Design Group at Technische Universität Ilmenau. This research position was financed by Deutsche Forschungsgemeinschaft (DFG) under the funding labels WE 1989/6-1 and WE 1989/6-3.

I would like to extend my heartily thanks to my doctoral supervisor Professor Dr.-Ing. Christian Weber for his support and supervision throughout my research work. He helped me to orient myself at the start of my research, provided constructive feedbacks on regular intervals, always spared time for research-oriented discussions and helped me in the organisation of technical equipment needed for the research. I would also like to thank him for his suggestions, corrections and clues for my scientific publications as well as my dissertation. Furthermore, I would like to extend my thanks to Professor Dr. phil. Heidi Krömker and Professor Dr.-Ing. Rainer Stark for being the evaluators of my doctoral thesis as well as for their suggestions and valuable feedback.

My heartily thanks also goes to my family, particularly my parents for their prayers, continuous support and patience in my time away from home.

Dr.-Ing. Stephan Husung accompanied me throughout my research as a mentor. I would like to thank him for the scientific discussions, his feedbacks about my scientific publications and constructive ideas. My big thanks also go to Dipl.-Ing. Andreas Liebal for proofreading my thesis, for the constructive ideas and discussions throughout my research. It was a pleasure to work with him. I thank my colleague Mr. Felix Röhnert for helping me in research equipment installations, Dr.-Ing. René Reich for constructive discussions about thesis writing and Mrs. Cindy Pietschmann for her assistance in organisational matters. I would also like to thank my colleagues Dr.-Ing. Gunhild Chilian, Dipl.-Ing. Antje Siegel, Dr.-Ing. Kersten Liebermann, Mr. Felix Harfensteller, Mr. Johannes Schleichert, Mr. Martin Petrich, Dipl.-Ing. Andreas Schmidt, Dr.-Ing. Steffen Lutz, Miss Regina Koreng and Dr.-Ing. Thorsten Brix for their support and encouragement during my research. I would also like to thank Dr.-Ing. Helge Drumm (technical supervisor of the FASP) for his technical support during my work in the FASP.

Ilmenau, November 2020

Atif Mahboob

Table of contents

List of figures

List of tables

Zusammenfassung

Digitale Methode und Modellen ermöglichen den Produktdesignern eine frühzeitige Evaluierung des Produkts, damit sie das Verhalten des Produkts und seine Interaktionen mit benachbarten Systemen in seinen späteren Lebensphasen besser verstehen können. Virtual Reality (VR) ist eine Technologie, die zum frühen Evaluierungsprozess beitragen kann, indem spätere Lebenssituationen eines Produkts schon in der Entwurfsphase angezeigt werden können. Die Anwendung von VR in der Industrie ist jedoch derzeit aufgrund des hohen Modellaufbereitungsaufwands und der limitierten Wiederverwendbarkeit vorhandener Modelle begrenzt. Daher befasst sich diese Arbeit mit der Entwicklung einer Methode, die die frühzeitige Evaluierung des Produkts innerhalb von VR und die Verwendung von VR im Produktentwicklungsprozess erleichtern kann. Diese Methode befasst sich mit dem Prozess der Entwicklung allgemeiner Verhaltensbeschreibungen zur Verwendung in VR, die auch wiederverwendet werden können, um dynamische Anwendungsfälle eines Produkts in den verschiedenen VR-Systemen abzubilden. Der Fokus liegt auf der Reduzierung des gesamten Aufbereitungsaufwands von VR-Modellen und auf das Verwirklichen einer hohen Wiederverwendbarkeit bereits vorhandener Modelle.

Die Kernkomponenten der Arbeit bestehen in der Verwendung von Model Based Systems Engineering (MBSE) zur Entwicklung allgemeingültiger Verhaltensmodellbeschreibungen, ihrer Verwendung beim Erstellen verschiedener Anwendungsfälle eines Produkts in einem VR-System und ihrer Wiederverwendung in den verschiedenen VR-Systemen. Die Systems Modeling Language (SysML) wird zur Beschreibung der Verhaltensmodelle verwendet, der Modellierungsprozess wird systematisch beschrieben und auch in Form allgemeiner Anwendungsrichtlinien für die spätere Verwendung zusammengefasst. Darüber hinaus wird eine dedizierte Physik-Engine verwendet, um die physikalischen Berechnungen für virtuelle Objekte in VR durchzuführen, welche auch mit SysML integriert ist. Diese SysML-Verhaltensmodelle zusammen mit der Physik-Engine bilden eine echtzeitfähige Produktanwendungssimulation in VR. Dieselben SysML-Verhaltensmodelle werden für verschiedene VR-Systeme verwendet, um Echtzeitsimulationen abzubilden und ihre Wiederverwendung zu validieren. Zwei VR-Prototypen wurden entwickelt, um die Wirksamkeit und Verwendung der vorgestellten Methoden zu demonstrieren. Schließlich wurde einer der Prototypen einer empirischen Untersuchung unterzogen, die mithilfe von Experten aus Wissenschaft und Industrie durchgeführt wurde.

Abstract

Digital methods and models help the product designers in performing early evaluations on a product that eventually help to gain understanding about a product's behaviour and its interactions with neighbouring systems in its later life-phases. Virtual Reality (VR) is a technology that can facilitate the early evaluation process by showing later life situations of a product as early as at the design stage. However, the application of VR in the industry is currently limited due to high model preparation effort and poor reusability of already prepared models. Therefore, this thesis pursues towards the development of a method that can facilitate the early evaluations of the product in VR and thus, facilitate the use of VR in the product development process. This method aims at achieving generic behavioural descriptions for use in VR that can be reused as well to form dynamic use cases of a product in different VR-systems. The focus lies on reducing the overall preparation effort of VR-models and on achieving high reusability of already created models.

The core components of the thesis consist of the use of Model Based Systems Engineering (MBSE) to develop generic behavioural model descriptions, their use in building different use cases of a product in one VR-system and their reuse in different VR-systems as well. The Systems Modeling Language (SysML) is used to describe the behavioural models, the modelling process is described systematically and is also summarized in the form of general-purpose guidelines for later use. Furthermore, a dedicated physics engine is used to perform the physical calculations on virtual objects in VR and is integrated with the SysML. These SysML behaviour models together with the physics engine are used to achieve a real-time product use case simulation inside VR. The same SysML behaviour models are used across different VR-systems to achieve real-time simulations and to validate their reuse. Two VR prototypes are developed to demonstrate the effectivity and use of the presented method. Finally, one of the prototypes is put to the empirical evaluation performed with the help of experts from academia as well as industry.

1 Introduction

It is the task of a product designer to consider all life-phases of a product at the design stage and to ensure that the finished product will fulfil the requirements arising from its different life-phases. General product life-phases according to VDI2221 [Vdi87] are product planning, product development, production planning, production, distribution, use, service and disposal. The products of today are becoming increasingly complex and at the same time, they are multi-disciplinary. The demands and expectations of customers are increasing and the overall development time is decreasing. One goal during the product development process is to gain an understanding of the product and its behaviour in its later life-phases so that the final solution complies with the needs of all the stakeholders. This compliance requires consideration of the needs of different stakeholders of the product at the product development stage. The use of different use cases of a product and their evaluation help to understand different situations that can arise in a product's later life-phases. However, many aspects of such an evaluation are highly dependent on the availability of a physical prototype of the product. The critical factors associated with the use of prototypes is the cost and availability in the initial design stages. Considering these challenges, there is an ever-increasing drift towards the use of digital models and methods in the product development process. These models and methods help the product designers to evaluate and to gain understanding about different aspects of a product and its behaviour. The Virtual Reality (VR) technology can serve as a major support tool, as it can help the designer by showing the product (visually and/or acoustically and/or tactile) and its behaviour in later life-phases. Thus, VR possesses great potential to facilitate the product development process by making an evaluation of the product possible as early as at the design stage.

The current methods that are deployed in the product development process mainly focus on the product alone and its functionality. A product cannot operate isolated on its own over its life span, instead, it comes in contact with different environments and actor(s). Actors are the persons with specific roles over the product's life cycle e.g. fabricator, assembler or the end-user. A product can require electricity, certain space, human interventions etc. for use or repair etc., therefore, it is of great importance to consider product's interactions with its neighbouring systems as well. Figure 1-1 shows different life-phases of a product with the respective actor(s). Every product in its different life-phases finds itself in different environments and also different human actor/actors come(s) in contact/interact with it. For example, during product production, a manufacturing machine operator is the actor and manufacturing environment is the environment of the product. Similarly, in any particular life-phase, the environment of the product and the human actor(s) can be referred to as the "context" of the product.

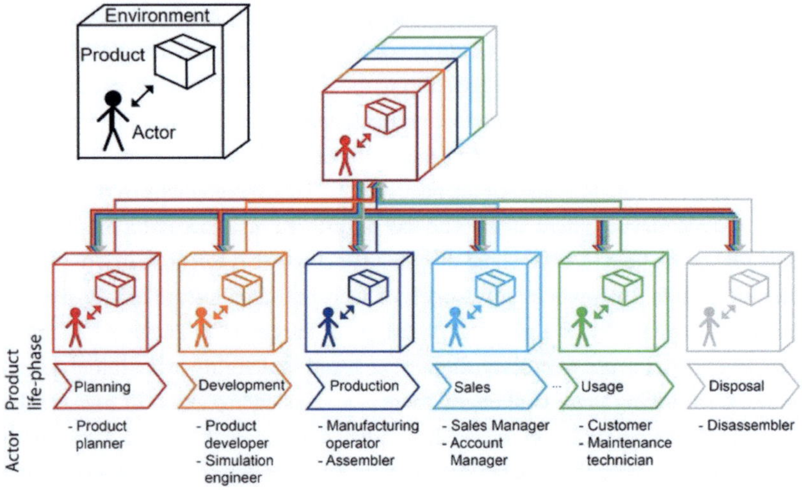

Figure 1-1: Context of product in different life-phases [Lie17]

Based on the different actor(s) and environments in different life-phases, the context of the product changes and so do the expectations from it. An evaluation of the context of the product can be achieved by building a contextual use case of a product in VR i.e. a product along with its environment and actor(s). Such a use case can facilitate the designer in foreseeing future interactions of a product by offering a later life-phase-specific situation in VR. The cost associated with the error corrections increases exponentially over the life span of the product [Ste04]. The earlier a design flaw is detected, the cheaper it is to resolve. Therefore, the early evaluation of the product's later life situations is vital for successful product development. The early evaluation of a product along with its context not only offer a better understanding about the product and its behaviour in later life, but it also possesses the potential of exploring new requirements of the product that may yet not have been considered during the design.

1.1 Motivation

Today the designer in the industry stays at the centre of the development process, where he/she has to either think of all the possible use cases of the product, or he/she uses different models and simulations to ensure fulfilment of product requirements. The product designer can be supported in the fulfilment of his/her tasks by means of a virtual model that can help the designer to efficiently configure different use cases of a product inside VR. One important aspect of this virtual model is the inclusion of an actor as the human interacting with the product in any particular situation. The designer will be the observer of this interaction instead of taking up the role of the actor. The main reason for the inclusion of an actor is the fact that the designer is usually the person with the highest knowledge about the product and thus, it can be very easy and intuitive for the designer to use/service the product. However, the same tasks may not be that intuitive for the actual product user or technician later. Therefore, it is advantageous to have an actor model in the form of a virtual human model in VR and let the product designer be the observer of the overall system. In the role of an observer, the product designer may not be directly manipulating the interaction, however, he/she can examine the interaction between the product and actor. The second important aspect of this virtual model is the inclusion of a product life-phase-specific environment. This means that the product is not being considered alone as the object of evaluation, instead, the environment of the product is also considered. The inclusion of an environmental model brings multiple evaluation aspects starting from spatial requirements to functional atmosphere into consideration. Furthermore, the interaction between the product and environmental model can also be evaluated.

Although VR offers great potential to facilitate the product development process in many ways, the use of VR technology is limited in the industry due to several reasons e.g.

- The time and the effort needed to prepare VR-models is high,
- the interoperability between different VR-systems is low and
- the available software solutions are highly application-specific.

The major developments in the field of VR are pushed by the gaming and entertainment industry. Therefore, the available VR solutions focus mainly on aspects such as the quality of visualisation, user-friendliness etc. The simulations/animations are prepared by programming the whole VR game/application in advance. The VR applications in the gaming and entertainment industry allow the in-advance planning of all the possible scenarios and thus, the advance programming of all application scenarios is possible. On the contrary, the VR applications in product development are based on design reviews and product evaluation scenarios. In such applications, the focus lies on aspects such as real-time simulation, behaviour building, model preparation effort, interoperability between different software tools and reusability etc. These aspects cannot be guaranteed by in advance programming methods. Furthermore, the in advance programming can bring bias in the evaluation process and the evaluation can lose its meaning.

Although the VR technology can offer a real-time simulation of product behaviour and thus makes it possible to perform a contextual evaluation of a product, its application is primarily limited to visualisation of product models. The behavioural simulations of the products are not largely performed inside the VR mainly due to the following limitation [Mah17b]:

- the great amount of time and the effort needed for the preparation of virtual models,
- poor/limited possibilities to reuse these models and/or parts of them and
- limited modification possibilities i.e. a small change can force a complete new preparation.

A reason for the high number of current visualisation applications can be identified as the relative ease associated with building a visual model of a product in VR. The geometrical product model that is normally available as a CAD model during product development can be imported into VR with the help of different geometrical exchange formats that are currently available. The currently available VR-software facilitate a wide range of import possibilities for these geometrical exchange formats [Lor16][Mah16]. However, there is no direct import or exchange method available for the import of behavioural simulation information (physical calculation incorporation, functional simulations etc.) in VR (see also 3.3). Furthermore, the different VR-systems (CAVE, Powerwall, HMD, sVR[1]) make use of different VR-software that are not compatible with each other. This means that

[1] sVR – Smartphone VR

an application developed for CAVE VR-systems does not work in HMDs. Considering the potential of VR and the challenges associated with its use in product development, there is a need for a new method that can integrate VR into the current product development process by addressing the challenges discussed here. Such a method can greatly facilitate the product designer in fulfilling his/her tasks.

1.2 Goal and scope of this thesis

The goal of this thesis is to develop a method that can facilitate an early evaluation of a product using VR technology. Figure 1-2 shows the overall model for product development in VR.

The VR-model consists of a product inside a life-phase specific environment in the presence of one or more human actor(s). The complete VR-model possesses an interface that allows the product designer (VR-user) to choose actor, product and environment models from a model database to build a contextual use case in VR and to observe the interplay of these in VR. VR technology refers to different VR-systems that are currently available.

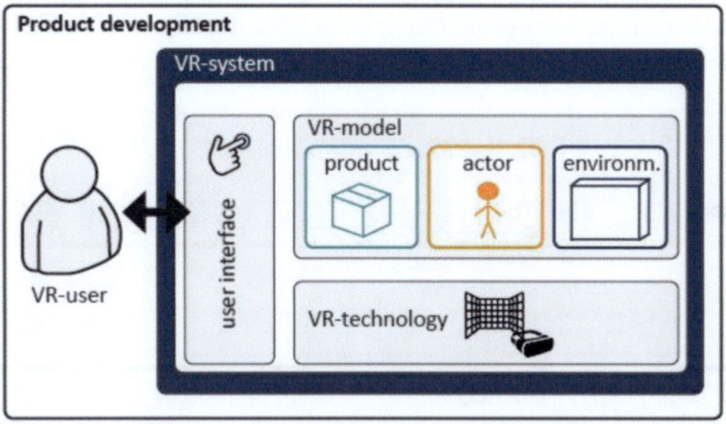

Figure 1-2: Product development in VR

The focus primarily lies on the development of a (generic) method for describing the behaviour of all included models i.e. product, actor and environment. The models of the product and the environment stay at the focus in this work and their behavioural models are also developed. The actor model is though also included as an important part in the methodological foundation, however, the implementation of a virtual actor model falls out of the scope of this thesis. Therefore, the actor model is only included in the methodological explanation and the implementation idea is not covered. Similarly, the user interface builds an integral part of the overall system but lies outside the scope of this thesis. The developed models of the product and the environment should achieve interoperability between different VR-systems and therefore, should also be tested in different VR-systems. Furthermore, these models should allow reusability so that different use cases can be built.

The developed method should help the product designer in foreseeing the interactions of a product with its environment and life-phase-specific actor(s) in its later life stages. Figure 1-3 shows a "user and task-oriented virtual model" containing the information about the actor (s) and environments from all life stages that should be made available at the development stage. Based on this information, different use cases can be built that may help the designer to observe the interactions of the product with its neighbouring systems (e.g. the actor(s) and the environment). The actor(s) and the environment in any particular life-phase are in the problem space and it may not be possible to directly influence them during the development process. However, the product itself being in the solution space can be evaluated and optimized to address the needs/requirements associated with the problem space at hand.

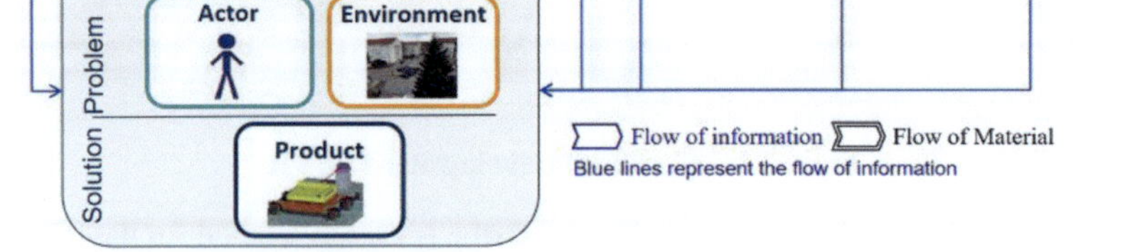

Figure 1-3: Information about the actor and the environment from product life-phases [Web16]

To attain the aforementioned goals following core areas are identified:

1. Integration of VR in the current product development process to enable early evaluation of a product
2. Development of a method to build the behaviour of a product along with the visual representations in VR while at the same time, reducing the effort needed to prepare VR-models
3. Generic description of VR-models with reusable contents to make different use cases of a product inside one particular VR-system and also in different VR-systems (CAVE, Powerwall, HMD, sVR) possible

After laying down the goals of the thesis in this section, the next section will provide the reader with a brief overview of the approach used in this thesis. Furthermore, the scope of the thesis and important methodological aspects will be briefly discussed.

1.3 Course of action

To reach the goals set in the last section, this work focuses on the development of a new method for the preparation of VR-models. After performing an analysis of currently available methods and challenges associated with the preparation of VR-models, a new method is presented that reduces the preparation effort by achieving reusability of the contents of VR-models. It is investigated on the methodological level to find the best approach for dividing the complete VR-model into reusable modules. This division allows the reusability of the contents of one VR-model in other models as well. The reusability of the contents of the VR-model is not limited to reuse within one VR-system, instead, the reusability perspective across different VR-systems is also discussed. To implement the reusable modules resulting from the methodological division of VR-model, the Model Based System Engineering (MBSE) approach will be put to test. As the goal of this work is to integrate VR in the current development process in the industry, the methods and models already accepted (to a certain extent) in the industry are analysed. An MBSE based approach is promising to pertain to the inclination of the industry towards the adoption of MBSE methods (detailed rationale in section 3.1). A concept for the integration of MBSE models to achieve an interactive simulation in VR and for the reuse of these models within one VR-system as well as the reuse across different

VR-systems are presented. Furthermore, a generic approach to incorporate physical calculations into VR is developed that can be used in different VR-systems irrespective of the VR-software in use.

To validate the method presented in this thesis, two prototypes are developed. The first contains a vacuum cleaner as a product and the second one contains 6 degrees of freedom (DoF) industrial robot as the product. Furthermore, the VR-model descriptions of these prototypes are used across different VR-systems to achieve example simulations. The vacuum cleaner application is subjected to empirical evaluations conducted with the help of academic and industrial experts. Having laid out the goal and course of action, the next section will provide the thesis outline and familiarise the reader with the structure of this thesis.

1.4 Thesis outline

A graphical demonstration of the contents of this thesis can be seen in figure 1-4. The first chapter introduces the research idea and its goals along with a brief introduction of the approach used in this work. The next chapter 2 talks about the fundamental models and methods relevant to the scope of this research. A general introduction of VR, its different systems and applications are presented. Also, it talks about important aspects of MBSE, development of models in MBSE and standardised languages available for the implementation of MBSE. Furthermore, a detailed explanation of the well-known modelling language in MBSE i.e. Systems Modeling Language (SysML) is provided and its meta-model is discussed.

Chapter 3 dives into the state of the art relevant to the use of VR and MBSE in the current product development process. The latest developments in the field of MBSE e.g. integration with the CAD system, integration with PLM system as well as the industrial pilot projects for MBSE incorporation and its use with VR are highlighted. This chapter also attempts to highlight the benefits of using VR in product development and its current application areas as well. An overview of the conventional ways of developing VR-models, their benefits as well as shortcomings and the justification for the need for a new generic method are presented. After analysing the currently available methods, this chapter identifies the research gaps and concludes on the formulation of concrete research questions.

After presenting the state of the art in chapter 3, chapter 4 talks about the methodology developed in the scope of this thesis. The requirements associated with the new method for descriptions of VR-models, division of complete VR-models in sub-models and modelling process in SysML are explained in detail. The modelling of structure, interaction and behaviour of sub-models is described systematically and the gained knowledge is summed up in the form of general-purpose guidelines. Furthermore, the use of sub-models to generate different use cases of a product is elaborated and reusability perspective of SysML models is discussed in details.

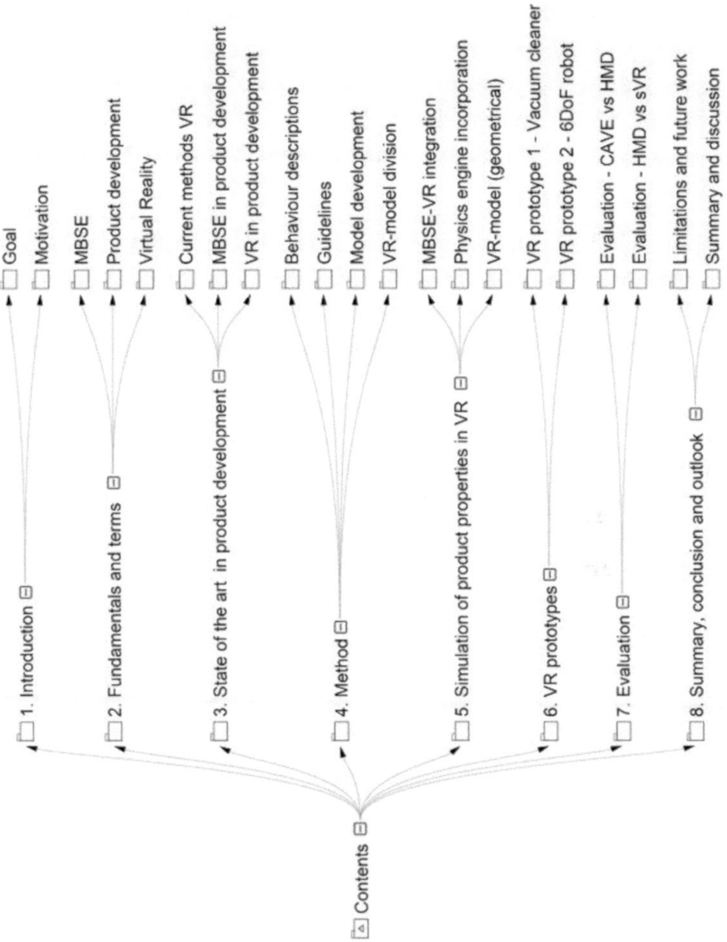

Figure 1-4: Structure of the thesis

Chapter 5 deals mainly with the VR side of the story and talks about the integration of SysML behaviour model with VR and also with the physics engine. The overall simulation process is elaborated here. It outlines the development of the visualisation model in VR, its integration with behavioural models from SysML and the integration of the physics engine in the VR simulation. As a result, a generic information exchange concept in-between different tools that are involved in the VR simulation is presented.

Chapter 6 examines the applicability of the developed method for use in different VR-systems and addresses the hardware-related aspects. In this regard, two VR prototypes are developed and their VR contents are reused to achieve VR simulations in different VR-systems. The first prototype is developed in all three VR-systems i.e. CAVE, HMD and sVR while the second one is developed inside HMD and sVR.

To collect industrial feedback about the developed applications, one of the VR prototypes was put to empirical evaluation. The method used to conduct the tests under this evaluation and the resulting outcomes are listed in chapter 7. Finally, chapter 8 summarizes the work and highlights the spots for future research.

2 Fundamentals and terms

This chapter provides an overview of the fundamental models and terms that will be later used to achieve the goals of this research. The focus here mainly lies on the renowned product development models/guidelines in place. Furthermore, it outlines the different VR technologies that include an overview of hardware setting, their capabilities and differences etc. This chapter does not attempt to cover all the models and methods in place in the above-mentioned fields, instead, it includes an overview of the ones that are most relevant to this research.

2.1 Product development process models

There are several models for product development that help in the product development process by providing basic development structures and approaches. Some of the important ones here are VDI2221 [Vdi87] for the design of mechanical products, VDI2206 [Vdi04] for mechatronics design, the System Lifecycle Management (SysLM) concept of [Eig16] and CPM/PDD approach of [Web05] for validation of product properties.

2.1.1 VDI2221 – Design of mechanical products

For the development of technical products that mainly focus on the field of mechanical engineering, the VDI2221 [Vdi87] is the general-purpose guideline. The different life-phases during the product life cycle according to VDI2221 are mentioned in figure 2-1. The needs of the market or the end-user or eventually own goals of a company drive the design process and lead to the definition of the task at hand. The design of the product takes place inside *Product development design* that is further concretised in the form of seven stages mentioned in figure 2-2. The first stage is the clarification of the task at hand by building the product specification or by extracting the expectations/requirements from the product/system. These specifications are analysed in detail to determine different underlying functions and the structural features in stage two. Once the functions are extracted, the exploration for possible solutions to achieve these functions starts and is handled in stage three. The result of this stage is the principle solution that may contain one or multiple solutions variants to achieve the desired functionality. Stage four analyses the principle solution and divides it into realisable modules. The result can be sub-systems or system elements that can be demonstrated in the form of sketches, graphs etc. The realisation of the shape of the system/ product takes place in stage five where decisive modules are developed. Focus is put on the completeness of geometry, materials and technical details so that further optimisation can be carried out. The result is preliminary design layouts of the system/product that are further detailed in stage six to move the design towards its final shape. Here segregation of elements may take place and different elements are grouped. The result of this stage is the overall design draft containing detailed information about the shape of the product described as sketches, components noted down as inventory lists and flow diagrams specifying the flow of information/materials.

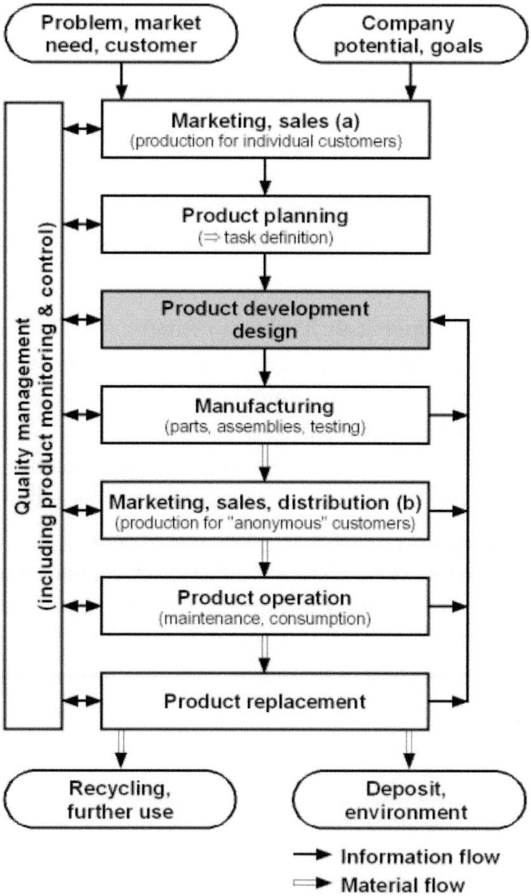

Figure 2-1: Product design as part of the life-phases of a product/system after VDI2221 ([Vdi86],[Vdi93] & [Vdi87])

The seventh is the last stage and is a preparation stage for the manufacturing phase. The product documentation involving its parts, sketches, inventory list etc. as well as the instruction manuals related to the product's later life stages are developed here. These lay down the procedures for manufacturing, assembling, testing, transportation and the final usage. An important aspect in figure 2-2 are arrows connecting towards the previous stages and the ones interconnecting all design stages. These indicate the possibility of iterating different stages to bring improvement to the design.

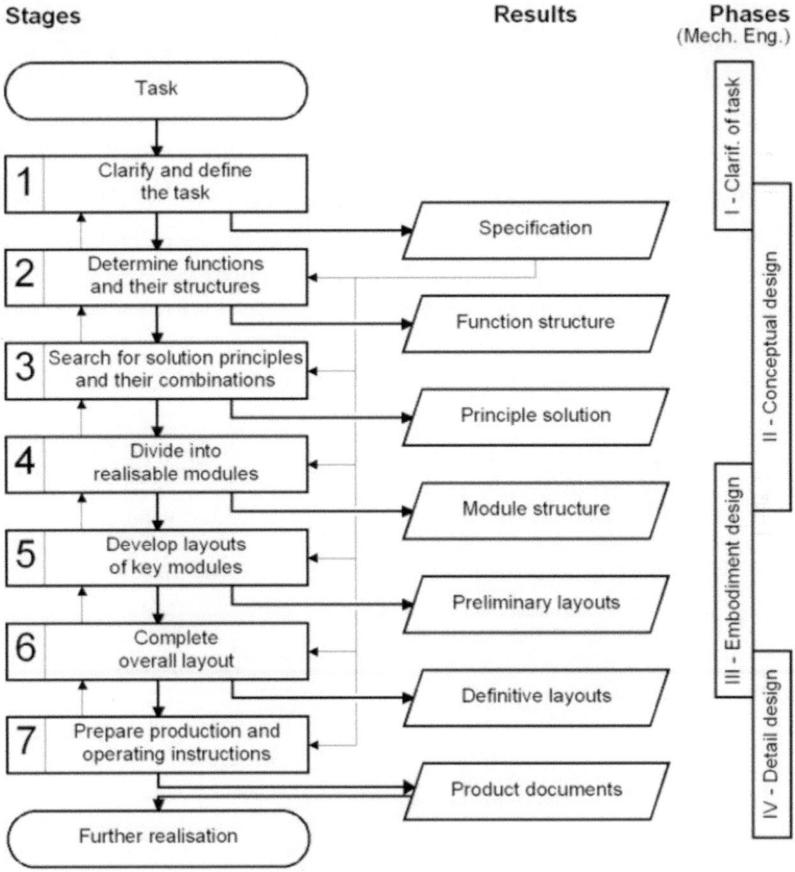

Figure 2-2: Steps during design according to VDI2221 ([Vdi86],[Vdi93] & [Vdi87])

Once the design stage is completed and the product documentation is prepared, the product enters the manufacturing phase. After manufacturing, the product enters the sales, marketing and distribution phase reaching eventually to the end-user. Finally, it enters its operation where it may need standard maintenance, repairing, servicing etc. At the end of its life cycle, its components may be recycled or disposed of eventually. An important aspect in figure 2-1 are arrows connecting later life-phases to the product development design phase representing the flow of information. This represents the availability of information about the product in its later life-phases that can eventually help to improve the next version of the product.

This thesis focuses on the design phase of the product and the developed method targets primarily the stages three to six in figure 2-2. This is done by ensuring the product properties with its context by means of a virtual model before moving towards completion of the development cycle and eventually to manufacturing.

2.1.2 VDI2206 – Design of mechatronics products

The design of mechatronic products is different from the design of mechanical products, as in the former case the products are interdisciplinary. VDI2206 [Vdi04] specifies the design guideline for mechatronic product and is intended to supplement VDI2221[Vdi87] and VDI2422 [Vdi94]. [Vdi04] presents the so-called V model for the development of mechatronic systems as shown in figure 2-3.

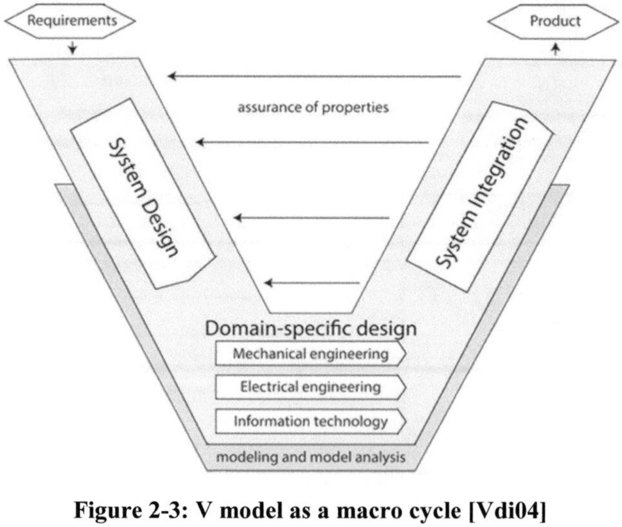

Figure 2-3: V model as a macro cycle [Vdi04]

According to the V model, the starting point of the development is the precise definition (as precise as possible) of the product's requirements. These requirements on one side define the needed product and on the other side build the measure for later assessment of the product design. As mechatronics systems are multidisciplinary systems, the system design aims to develop the solution concept for the final product. Here the breakdown of the overall system functionality takes place and sub-functions are defined. Furthermore, different solution elements for these sub-functions are explored and their performance is tested. The solution elements are separated into the respective domain in domain-specific design. The corresponding elements are detailed and their performance is ensured by performing different calculations. In system integration, an overall system is formed by joining the solutions from domain-specific development. It is important to check that the design is aligned with the solution concept and the requirements, and this is achieved by performing assurance of properties. At the end of system integration, a product is developed. There are several extensions of the V model already performed in different researches in different disciplines. One of these extensions for use in the development of cyber physical systems is elaborated in the next section 2.1.3.

2.1.3 System Lifecycle Management

The development of multi-disciplinary products involves the use of different authoring tools/software. To have consistent development, the information present in different tools should be linked with each other in some manner to achieve traceability. System Lifecycle Management (SysLM) [Eig16] is an information management model for use in product development. It builds on the Product Lifecycle Management (PLM) solutions in place today and extends them to achieve an integration between different authoring tools. This integration is expected to be implemented along the system life cycle, where the system models in all development phases are connected with one and another to achieve traceability. The emphasis is put on the use of a model-based approach instead of document-based approaches. [Vdi04] lays down a systematic approach for the development of mechatronics systems. However, it does not address all aspects related to model-based design & the development of cyber physical systems. Therefore, the original V model is further extended under the SysLM concept to address the aspects related to model-based development for complex systems. The extended V model is shown in figure 2-4.

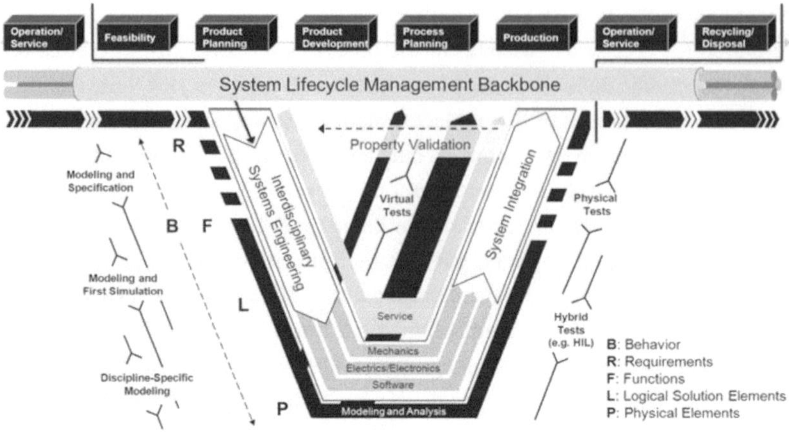

Figure 2-4: Extended V model for Multi-Disciplinary Product Development [Eig16]

The extended V model talks about the early design (left-wing) as well as the detailed design aspects. The left side of 'V' deals with the model-based description of the system under development in its early phases and mainly describes the specifications, first simulation and the discipline-specific modelling of the system. *Modelling and Specification, Modelling and First Simulation & Discipline-Specific Modelling* are three aspects of the system and their overlapping is further represented as requirements (R), functions (F), logical solution elements (L) and physical parts (P) using the RFLP-approach [Kle13]. The modelling of RFLP is performed by using different languages and authoring tools. The goal of this extension of the V model is to achieve vertical as well as horizontal traceability and it can be achieved by semantically linking the different model elements. The implementation of this linking can be achieved by using system models created using SysML, as it allows the description of system elements, the links between them as well as the integration and management of information in the context of SysLM concept. After mentioning the domain-specific design on the bottom, the right-wing of the extended V model talks about the system integration. The validation of properties can be performed by conducting physical tests on the actual product (physical test) or hybrid tests or by means of fully virtual tests. The SysLM concept of [Eig16] and the developed extension of the V model rightly points out that using MBSE and performing early virtual tests can enable early validation of product properties. VR technologies can be very helpful for conducting these virtual tests. As a matter of fact, the virtual tests can be performed much earlier than the physical tests and thus, the extended V model is smaller when using virtual tests.

2.1.4 CPM/PDD

One approach that is very relevant to the work presented in this thesis is the Characteristics-Properties Modelling (CPM) and Properties Driven Development (PDD) approach for the assurance of product properties [Web05]. This approach establishes a clear distinction between the characteristics and the properties of a product. The characteristics (figure 2-5 left) here are the structure and the constituents of a product and the properties (figure 2-5 right) are the product's behaviour.

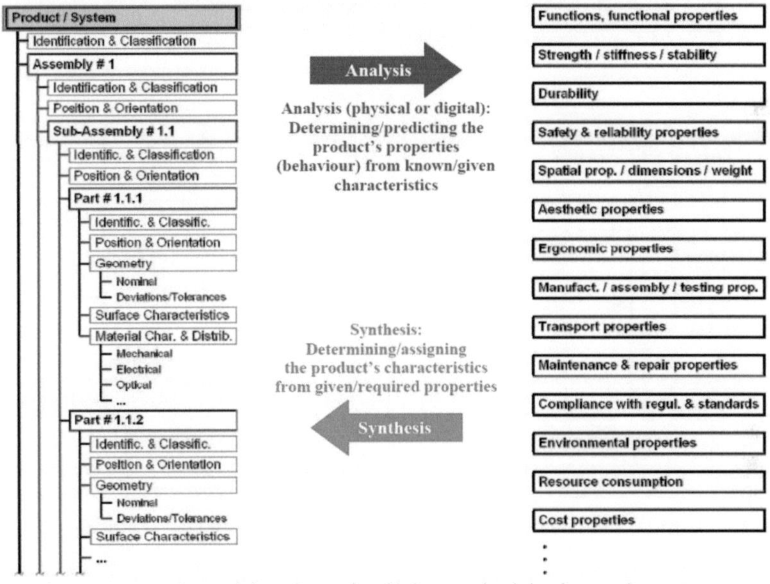

Figure 2-5: Characteristics (left) and properties (right) with their relations [Web05]

The determination of product properties can be performed by using the appropriate method/tools while considering the external conditions. These external conditions may consist of the assumptions or simplifications to perform the calculation of properties or external influences. Thus, the calculated properties are only valid against the considered external conditions. In this approach, the product development process is seen as cycles of synthesis and analysis steps. Synthesis is the process of determining product characteristics for the given properties or the properties that have to be achieved at the end. The analysis is the opposite of synthesis process i.e. determining product properties for the given characteristics. Before looking at the synthesis and analysis steps in details, it is important to highlight the important parameters mentioned in the original approach, as they are needed to understand the basic model of synthesis and analysis (see figure 2-6).

- **C_m / C_i**: Characteristics ("Merkmale" in German)
- **P_n / P_j**: Properties ("Eigenschaften" in German)
- **PR_n / PR_j**: Required Properties
- **R_n / R_j**: Relation between characteristics and properties
- **EC_n / EC_j**: External conditions

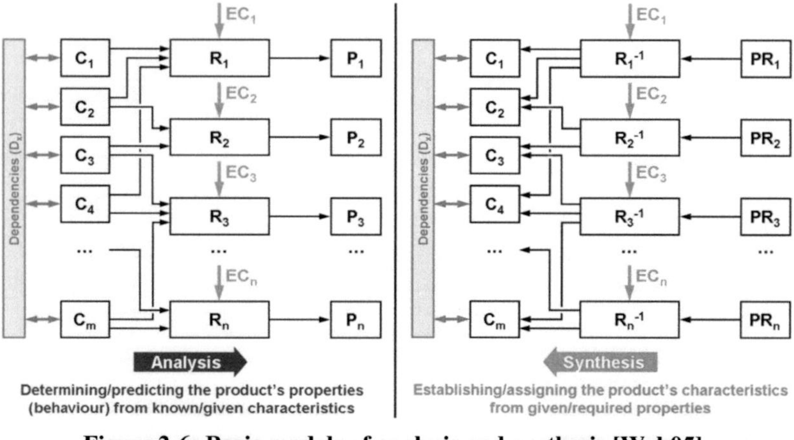

Figure 2-6: Basic models of analysis and synthesis [Web05]

The analysis step takes the characteristics **(C_i)** and calculates the properties/ as-is properties **(P_j)** of a product using the **(R_j)** that define the relationship between the former two. **(R_j)** are in the form of matrices that may contain equations and these equations define the relationship between a certain property and the characteristics. However, during the product development process, when the finished product is not available yet, the properties can only be predicted against the given characteristics. This can be achieved by using available digital models and methods. The synthesis step is formally the opposite of analysis. (**PR_j**) are the required properties or the final goal of the development process that the designers want to achieve. The synthesis step starts with the required properties **(PR_j)** and tries to determine the product characteristics by considering the relation between the two. However, the relation between the two here will be inverse of the relationship in the case of analysis. Therefore, the relationship in the case of synthesis is **(R_j^{-1})** that can be obtained by taking the inverse of **(R_j)** that is the relationship matrix. Although the inversion of the relationship looks like an easy solution, it is not possible in all the cases. The explanation for this phenomena is the simple fact that not in all the cases, it is possible to have an inverse solution for a matrix. In this way, the use of the basic models presented in figure 2-6 support the product development process. An extended scheme of the product development consisting of cycles of synthesis and analysis process is present in [Web11] and shown in figure 2-7.

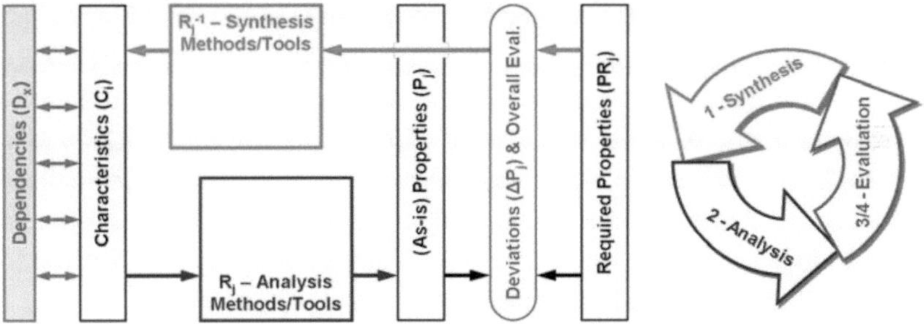

Figure 2-7: Scheme of the product development/design process consisting of cycles of synthesis-analysis-evaluation [Web11]

The incorporation of evaluation step completes the development circle. The results of the analysis step provide the as-is properties **(P_j)** that are valid against the given characteristics under consideration of the external conditions. Now, these as-is properties **(P_j)** can be compared with the required properties (**PR_j**). The difference/deviation (**ΔP_j**) is the difference between as-is and required properties. This difference is the driver of the development process, as it may represent the discrepancies in the current design and based on this the designers can perform an overall evaluation to decide about how to proceed further. In this way, the CPM/PDD can be used in a cyclic way using different methods/tools to ensure product properties.

(**EC_j**) are the external conditions and may contain the context of the product (defined in chapter 1) that has to be considered and a change in the context can change the calculation results. For instance, the external disturbances, influences, material flows etc. may exist in external conditions, as during the product's operation, they may originate from interacting actor(s) or the environment or other neighbouring systems. Therefore, the context of the product i.e. life-phase specific actor(s) and environment can be seen as a part of external conditions considered in CPM/PDD approach. Hence, the consideration of external conditions during the calculation of product properties makes this CPM/PDD approach very relevant to this thesis. These external conditions are extended in the scope of this thesis to explicitly include actor(s) and environment (details to follow in sub-section 4.1.2).

2.1.5 Summary

Section 2.1 provides an overview of the product development process models most relevant to this thesis. VDI2221 provides general guidelines for the development of mechanical systems and the tasks involved in this process. The VR-model presented in this thesis should be available at *"Product development design"* (in figure 2-1) stage to support the designer. This VR-model shall not be limited to the evaluation of mechanical products and thus, it is important to have a look at the development process model for an interdisciplinary product e.g. VDI2206 from section 2.1.2. The V model is very important as it lays down the guidelines for the development of mechatronics systems with multi-disciplinary design aspects. To form a more generic application also beyond mechatronics systems, the SysLM concept extends the V model to present a general-purpose guideline while using a model-based approach. Furthermore, the concept of integration of multiple discipline-specific tools inside one model-based development approach helps to achieve traceability and also plays a vital role in the development of the methodology presented in this thesis. The evaluation of a product alone can be misleading as the external conditions e.g. environmental condition, assumption during design etc. can largely influence the product performance. The CPM/PDD model is relevant to this thesis as it incorporates the external conditions and states that the validity of calculated properties is subjected to the given external condition. Such a consideration of external factors can be considered an analogy to the context of a product as mentioned in chapter 1. Considering its relevance, the CPM/PDD model will be used and extended in the scope of this thesis and this extension is presented in chapter 4.

The next section will introduce the reader with VR and provide a detailed overview of its hardware as well as a few of its general applications in product development.

2.2 Virtual Reality

Considering the benefits of Virtual Reality (VR) in product development, it is identified as a core part of this work. Therefore, before diving into the state of the art, this section provides the reader with a detailed introduction about VR, its applications and hardware technologies.

2.2.1 What is VR?

There is no standard definition of VR available, however, it can be understood as a presentation and perception of an intended reality that is implemented by real-time computer-generated virtual environments. In most scientific papers and in practice, VR is limited to graphic representations and visualisation although other human senses (e.g. haptics, acoustics) may be integrated as well – and are sometimes quite important, especially in the case of product development. Another definition of VR by [Bis92] is

"*real-time interactive graphics with three-dimensional models, combined with a display technology that gives the user the immersion in the model world and direct manipulation*"

As the term is self-explanatory, VR talks about a reality that is virtual or does not exist. A virtual surrounding built with the help of computer-generated graphics, presented as 3D representations along with object-level interaction provides the user with the feeling of being inside the virtual surroundings.

The key elements [She02] of a VR are

- a virtual world,
- immersion,
- sensory feedback
- and interactivity

The virtual world is an imaginary space or a description of a collection of objects in space and immersion refers to the immersion into an alternate reality or point of view. The sensory feedback is the feedback from VR e.g. based on user's physical position and one form of interactivity may be the ability to affect computer-based world [She02]. A VR-scene consists of different objects/surroundings that are three dimensional and the experience is different from normal computer screens as the depth information can also be perceived. The objects in a VR-scene are basically computer-generated graphics that are designed in advance. Immersion is defined as the feeling of being inside a 3D environment that can make the user unaware of his/her actual surroundings. In a physical world, humans interact with objects by touching, grabbing or manipulating them. In VR, the interaction can be achieved in multiple ways e.g. using interaction devices, voice commands, gestures etc. Some of the very basic interaction techniques for VR are presented in [Min95]. Although there are already several VR interaction devices and methods available, this topic remains the focus of continuous development (also see [Lav17], [Bow06] and [Bow08]). One very crucial requirement from VR technology is that it should possess real-time conditions, only then can an experience comparable with reality be obtained. Although VR has stayed in the focus of research for the last 40 years, it has only been recently made possible that the applications of VR are reaching domestic consumers. The developments in computer hardware have contributed mainly to make VR inexpensive and real-time at the same time.

Figure 2-8 shows the main components of a VR-system. VR comprises of two main parts i.e. hardware and the software [Bam13]. The hardware consists of the main computer for computations, displays for presentation of the graphics to the user, interaction devices and the tracking system to track user movements. The software side consists of the software tools and the databases that may contain already developed VR contents. The software allows the creation of VR contents in the form of a scene and performs the rendering of these contents during the simulation.

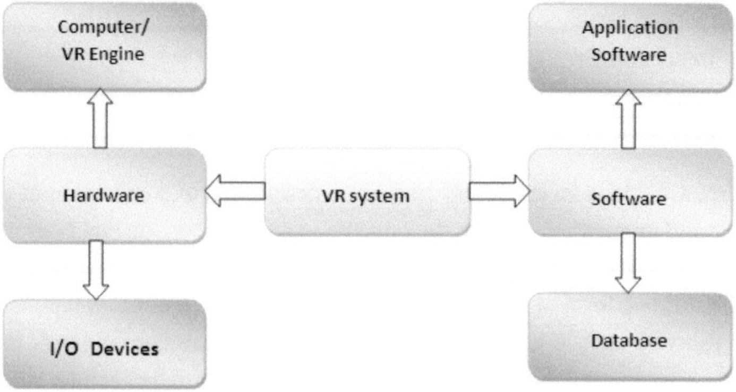

Figure 2-8: Components of VR [Bam13]

VR has found its application in diverse fields of interest. Some of the very important application areas of VR are in

- the entertainment industry e.g. 3D cinema and gaming,
- education,
- medicine,
- the aviation and automobile industry
- and in training.

A further detailed overview of VR applications will be provided in sub-section 2.2.4, sections 3.2 & 3.3.

2.2.2 Overview of different VR-systems

The last sub-section provides the general introduction of VR and this sub-section provides an in-depth look at different VR-systems (hardware systems). The main VR platforms that are used currently in industry and academia are:

- CAVE – Audio-Visual Automatic Virtual Environment [Cru92] and Powerwall
- HMD – Head Mounted Display [Mel01]
- Smartphone VR (sVR)

2.2.2.1 Cave Automatic Virtual Reality and Powerwall

CAVE is a multi-sided immersive virtual environment setup that can contain between 3 to 6 walls. A 6 sided CAVE builds a complete virtual room with four virtual walls, a virtual roof and a virtual floor. 3D contents are projected onto the walls and using the polarized glasses the user can experience immersive virtual reality. The glasses help to show each eye of the user different imagery and thus helps to create the illusionary 3D world. CAVE can also incorporate loudspeakers to provide directional sound. Figure 2-9 shows the flexible audio-visual stereoscopic projection system available at Technische Universität Ilmenau. It consists of three walls that are powered by two projectors each. Each visualisation projector has its own dedicated computer to process the projection information.

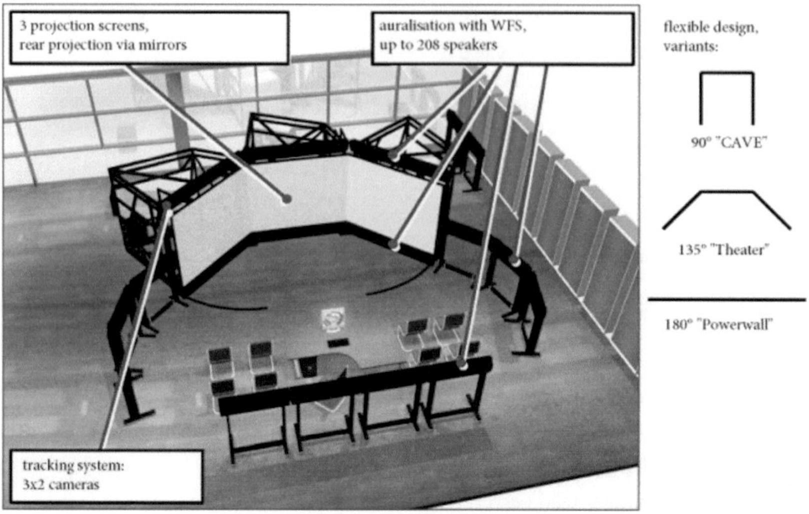

Figure 2-9: Flexible audio-visual stereoscopic projection system (FASP) [Hus14]

There are 208 speakers mounted on and around this system that are controlled by the WFS [Bra04] system (wave field synthesis system). Each projection wall has two tracking cameras to track the position of the user's head and the interaction device. The complete tracking system is controlled by a separate computer. An additional computer is used as a master computer to create VR content and to perform the rendering of this content onto the walls. This system is flexible in its design and can be used in three different configurations i.e. CAVE, Theatre, Powerwall. The flexible visual projection system (audio excluded) is used in this thesis to develop the example VR application for CAVE type VR-system (see section 6.1.1).

Powerwall itself is a cheaper version of the CAVE type system, however, it is normally based on a single wall with projectors attached to present 3D content. Similar to the CAVE system, Powerwall also possesses its own tracking system and interaction devices.

2.2.2.2 Head Mounted Display

The Head Mounted Display (HMD) on the contrary to the CAVE type system, projects the imagery directly on the retina of the human eye. The position tracking is obtained by additional sensory external to the HMD. There are several HMDs available in the market from different manufacturers. HTC Vive[2] can be taken as an example of the overall system to gain understanding about the HMDs. Figure 2-10 shows an example hardware setup for HTC Vive. It contains the HMD that projects a different image on each eye and includes the sensory system that can be detected by tracking cameras.

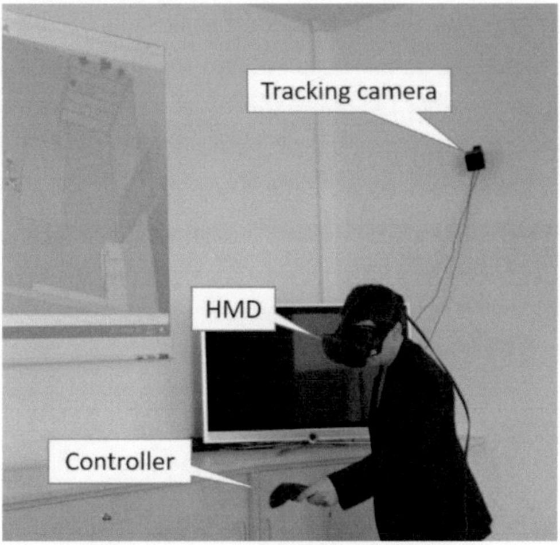

Figure 2-10: Head Mounted Display (HMD) [Mah18b]

HTC Vive comes by default with two tracking cameras and two controllers that serve as the interaction devices. The tracking cameras are responsible for the tracking of the head movements of the user by detecting the position and orientation of the HMD and also for the tracking of the interaction controller. One computer with high computation power and high-end graphics card is needed to render the VR contents in HMD. Furthermore, there are additional peripheries available to track the complete body movement, eye movement etc. of the user that are not discussed here. Similarly, the complete VR hardware for HMD (even for other HMDs that are available in the market) contains a high-performance computer for rendering, a tracking system, controllers and the HMD itself.

[2] HTC Vive: https://www.vive.com/de/ [last accessed on 09.03.2020]

2.2.2.3 Smartphone VR

Another relatively new technological development in the field of smartphones is their capability to render small (not extremely computation demanding) VR applications. The basic requirement for having VR experience with a smartphone is to have a headset/holder to place the smartphone and the smartphone itself should possess a gyroscope sensor. The gyroscope is used to keep track of the head movement of the VR-user. The basic principle of rendering the VR content onto the smartphone is to divide the phone screen in two and show two different images on each side. The headset/holder facilitates this by allowing each eye to see different imagery and blocks the unnecessary light from entering the eye. There are several headsets/holders available for smartphone VR (sVR), for simplicity Google Cardboard[Mac15] can be taken as an example which can be seen in figure 2-11.

Figure 2-11: Google Cardboard[3]

The main limitation of the smartphone VR is the absence of a positional tracking system. Although the gyroscope can detect the head movements (orientation only) of the user, the position tracking is not directly possible using the smartphone. There are methods available for indirect positional tracking implementation in a smartphone, however, they rely on additional sensory modules (e.g. see [Bos13] and [Fan17]). The accuracy and effectiveness of these modules is a separate debate that is not included here to keep the research effort focused on the goals of this thesis. Therefore, smartphone VR applications do not offer (by default) the position change in the VR-scene against the changes in the user's position. Due to the absence of a direct positional tracking solution, the current application of sVR is

[3] Google Cardboard: https://vr.google.com/cardboard/ [last accessed on 09.03.2020]

primarily in the visualisation of 360° videos as can be seen in [Bor17]. To facilitate the use of sVR in the product development process, a concept for achieving positional tracking in sVR is developed under the scope of this thesis. The implementation of this positional tracking along with the discussion about its preciseness is presented in 6.1.3.2.

The level of immersion or sense of presence in sVR compared with dedicated HMDs is also found to be of comparable quality ([Ami16] and [Pap17]). An idea of direct implementation of positional tracking in sVR can facilitate its use in industrial VR applications. Furthermore, there are critical factors associated with the use of sVR since the quality of the VR experience is highly dependent on the depth of pixels per inch (dpi) and the processing power of the smartphone. Both of these factors are directly dependent on the technological implementations in smartphones and are expected to improve with technological advancements in future.

2.2.2.4 Summary

Sub-section 2.2.2 has provided an overview of the four major VR-systems, their hardware setup and important components. All these VR-systems differ considerably from each other based on their application area, cost, installation requirement etc. Without going into the detail of the actual current cost of VR-systems, it can be concluded with fair ease that CAVE and Powerwall are the more expensive VR-systems mainly used by industrial OEMs. They can facilitate a multi-user visual experience, however, the point of view and interaction possibilities are mainly optimised for only one user. Furthermore, they require fixed installation space and hence, offer no mobility. HMDs are cost-effective VR solutions that lie in affordable cost range for small and medium-sized enterprises as well as for domestic users. They are optimised for one user's experience and hence, do not support multi-user visualisation. Although they also need a fixed installation space, they can be easily transported due to the limited needed hardware and hence offer a mobile VR-system. sVR is lowest in terms of cost and needed hardware. However, the low display quality, absent positional tracking and the limited processing power are the main limitations of this VR-system. Pertaining to very limited hardware setup (i.e. smartphone, headset/holder and interaction device), they are most suitable concerning the mobility and also do not require a fixed installation space.

Such a mobile VR-system may offer great help to industrial companies for making product presentation at remote locations.

2.2.3 Physics engine integration in VR

A VR simulation without physical calculation has limited meaning for an application in product development. Therefore, this section will have a look at the possibilities of performing physical calculations on virtual objects in VR. Performing the physical calculations on the objects in a VR-scene is purely the software related capability. Looking at the VR-software that are available today in the market, it is possible to do a clear division based on VR-system they are compatible with (see table 2). For instance, the CAVE type VR-system mainly works with commercial rendering software like RTT Deltagen[4] from 3DEXITE, IC.IDO[5], Visionary Render[6] etc. On the other hand, the well-established VR-software tools for HMD and sVR are mainly the game development engines like Unity3d[7], Unreal Engine[8]etc. The VR-software for CAVE type VR-systems focus on visualisation primarily and they either offer no direct physics calculation module or if available, its use is based on programming/visual scripting of the individual objects inside the VR-software. Unity3d and Unreal Engine implement the physics calculation modules in the software package in the form of libraries that can be used to add physics computations for individual objects. The problem with these solutions is the extensive programming needed to implement the physics calculation on the objects. Furthermore, the VR-software solutions available are not very well standardised. For example, VR content transfer from one to the other VR-software hardly enables the geometrical surface transfer and the simulation/animation exchange is an even more difficult process. The reason for this limitation is the fact that the language, syntax and implementation logic for simulation is different in each VR-software. There are several open-source physics engines available that can be used in VR. An evaluation of these physics engines for use in assembly simulation is presented by [Hum12]. As this thesis aims to develop a generic description method

[4] RTT Deltagen: https://www.3dexcite.com/de/software/deltagen/ [last accessed on 09.03.2020]
[5] IC.IDO : https://www.esi-group.com/de/software-loesungen/virtual-reality/icido [last accessed on 09.03.2020]
[6] Visionary Render : https://www.virtalis.com/visionary-render/ [last accessed on 09.03.2020]
[7] Unity3d: https://unity3d.com/de [last accessed on 09.03.2020]
[8] Unreal Engine : https://www.unrealengine.com/ [last accessed on 09.03.2020]

of a VR-model applicable in different VR-systems, there is a need for a more flexible solution for the incorporation of physics calculations in VR. The incorporation of physical calculation in the newly developed method is based on external physics engine integration in VR to make it VR-software independent. The detailed discussion about this topic is included in section 5.2.

2.2.4 VR beyond visualisation/extended VR application

In addition to the visual representation of the product and its behaviour in VR, the haptic interaction and the simulation of acoustical properties of the product are also the topics of great interest. However, a detailed discussion about haptic interaction devices and acoustical simulations is not included in this thesis. For reader's discretion, a few of the researches addressing these topics are mentioned in this sub-section.

A virtual acoustic model for simulation of the noise generated by a passing vehicle is developed by [Sie16]. This model creates individual noise sources for different components in the car e.g. tires, the engine and exhaust system etc. Each sound source has its own directional characteristics and the environmental conditions are simulated by sound reflections. This application is very helpful to examine the acoustical requirements related to a vehicle. There are more similar audio-visual applications in VR for the simulation of vehicle noise e.g. [Alb14][Hus14][Hus15] [Sie17a][Sie17b]. As changed acoustical characteristics may be a warning for a malfunctioning product, acoustical simulation analyses of a product can help analyse critical product functions. [Hus14] develops an acoustical application in VR while using the wave field synthesis [Bra04]. Individually recorded sound sources are used inside VR to simulate the acoustical behaviour of a product. A similar application of a pick and place machine acoustical simulation is developed in [Hus11]. To do such an analysis, the physical behaviour of the product has to be modelled first.

Similar to acoustical simulation, the incorporation of haptics is an important feedback channel in VR applications and therefore, stays an important research area in VR applications. Several haptic interaction devices are already available in the market for different VR-systems. A more realistic haptic feedback leads to a more realistic perception and more immersion for the VR-user.

2.2.5 Summary

This chapter has so far dealt with the general models for product development relevant to this research and their relevance is brought into focus. A general overview of different VR-systems and their hardware settings is provided as well. As the developed method uses MBSE, it is important to have a general introduction of MBSE, its origin and its implementation languages. These aspects are covered in the next section.

2.3 Systems Engineering

Model based Systems Engineering (MBSE) is a sub-branch of Systems Engineering (SE) and therefore, it is important to first introduce SE briefly before discussing MBSE. Systems Engineering (SE) focuses on the development of interdisciplinary and complex systems. It uses design thinking to capture the needs of all the stakeholders of a system and its functionality at the early stages of development. Starting from the documentation of the requirements throughout the development process, it focuses on the management of information, all necessary artefacts of a system and also the consistency of the system interfaces. As the products of today are increasingly gaining complexity, the SE approaches have gained significant interest in the last years. As a result, several approaches and models have been developed to facilitate and implement systems thinking inside the product development process. There are several definitions of SE currently available, a precise one is given by *The International Council on Systems Engineering* (INCOSE) [Inc20] as:

"Systems Engineering is an interdisciplinary approach and means to enable the realization of successful systems. It focusses on defining customer needs and required functionality early in the development cycle, documenting requirements, and then proceeding with design synthesis and system validation while considering the complete problem: operations, cost and schedule, performance, training and support, test, manufacturing, and disposal.

Systems Engineering integrates all the disciplines and specialty groups into a team effort forming a structured development process that proceeds from concept to production to operation. Systems engineering considers both the business and the technical need of all customers with the goal of providing a quality product that meets the user needs." [Inc20]

SE comprises of three building blocks i.e. system architecture, system requirements and system behaviour [Alt12]. System architecture defines the structure and components of the systems. The definition of internal interfaces between the system components as well as between sub-systems and external systems are also defined here. Systems requirements consist of functional as well as the non-functional requirements of the systems. System behaviour contains the behavioural definitions of the system and its components that can be used to perform various tests on the system under development, generate different configurations or generate executable code etc.

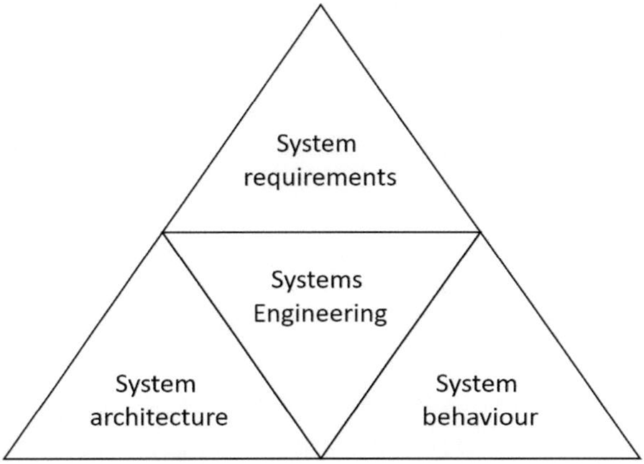

Figure 2-12: Three building blocks of Systems Engineering [Alt12]

2.3.1 Model Based Systems Engineering (MBSE)

The standard SE approaches use documents to manage the information and knowledge about the system. These documents usually contain spreadsheets, drawings and written specifications. However, the management of these documents can be prone to error and be time-consuming which make them a difficult task to handle when the systems under development are complex and relatively large. To tackle this problem, MBSE is a sub-branch of SE that emphasizes on the use of models instead of documents during the development process. MBSE as per [Inc07] is defined as :

"Model based systems engineering (MBSE) is the formalized application of modeling to support system requirements, design, analysis, verification and validation

activities beginning in the conceptual design phase and continuing throughout development and later life cycle phases."

The knowledge about the system is organised inside models usually defined using graphical modelling languages. These models allow the capturing of the requirements, structure and behaviour of the complete system and can be used to perform early validations on the design. The use of models helps to attain consistency throughout the development process by strengthening the communication between development teams and also helps to perform early verification and validation activities.

2.3.2 MBSE implementation

The implementation of MBSE inside an organisation or a development team involves three main components i.e. method, tool and language [Har15]. A method is needed to perform the needed modelling inside a modelling tool using a modelling language (possibly with standardised specification).

Among the already established methods in MBSE [Est07], INCOSE *Object-Oriented Systems Engineering Method* (OOSEM) and *The Systems Modeling Toolbox* (SYSMOD) modelling approach are two significant ones. OOSEM [Pea12] is a tool- and vendor-neutral approach that integrates a top-down approach with a model-based approach. It uses SysML and object-oriented concepts to support system requirements and context definition, system design, analysis and verification. SYSMOD [Wei15] focus on a user-oriented approach for carrying out requirement engineering and for developing system architectures. It is independent of the tool vendors and provides guidelines for performing different process activities.

The modelling tool in MBSE plays an important part in the development and validation of the system under development. Ease of modelling, compliance with specifications and standards, offered functionality, cost etc. are few of the factors that affect the choice of such a tool. The tools that use SysML specification for

MBSE implementation are available both commercially and in open source versions. *Cameo Systems Modeler*[9], *Rational Rhapsody Designer*[10], *Windchill Modeler*[11], *Enterprise Architect*[12] and *Visual Paradigm*[13] are among few of the commercial available MBSE tools. *Eclipse Papyrus*[14] and *Modelio*[15] are among open-source modelling environments.

Unified Modeling Language (UML) [Omg17b] and Systems Modeling Language (SysML) [Omg17a] are two graphical languages in MBSE. UML is widely accepted around the world for the development of software systems. The perception of UML being a description language for software engineering pushed the development of its more abstract version i.e. SysML. SysML uses a big sub-set of already developed UML specification 2.0 along with some extensions to form a more general-purpose graphical modelling language. Both SysML and UML are non-executable languages, however, they support model execution through translations/ transformations into executable languages or Petri nets [Des01]. The modelled behaviour in UML/SysML can be executed using languages such as Modelica [Mod17] [Fri98], Simulink [Ang11], VHDL-AMS [Iee17] etc. Almost all the existing commercial tools have implemented the execution of models by the integration of execution logic inside the modelling tool. As a result, the models can now be directly executed inside the modelling tools. A more extensive overview of the meta-model of SysML is provided in the next section.

2.4 Systems Modeling Language (SysML)

SysML is a graphical modelling language that is an extension of the existing UML specification. SysML evolved as a result of the combined effort of *Object Management Group* (OMG) and *International Council on Systems Engineering* (INCOSE). It uses a large subset of UML, includes modified as well as new elements

---

[9] Camero Systems Modeller: https://www.nomagic.com/products/cameo-systems-modeler [last accessed on 09.03.2020]
[10] IBM Rational Rhapsody: https://www.willert.de/software-tools/modellgetriebene-softwareentwicklung/ibm-rational-rhapsody/ [last accessed on 09.03.2020]
[11] PTC Windchill Modeler (Integrity Modeler): https://www.ptc.com/de/products/plm/plm-products/windchill/modeler [last accessed on 09.03.2020]
[12] Sparx Systems – Enterprise Architect: https://www.sparxsystems.de/ [last accessed on 09.03.2020]
[13] Visual Paradigm: https://www.visual-paradigm.com/ [last accessed on 09.03.2020]
[14] Eclipse Papyrus: https://www.eclipse.org/papyrus/ [last accessed on 09.03.2020]
[15] Modelio: https://www.modelio.org/ [last accessed on 09.03.2020]

and diagrams. It uses graphical diagrams to model a system under development and can facilitate analysis, verification and validation activities on the design. These diagrams are categorized into structure, behaviour and requirement diagrams. Figure 2-13 shows an overview of the SysML diagram from the current specification i.e. version 1.6 [Omg18b].

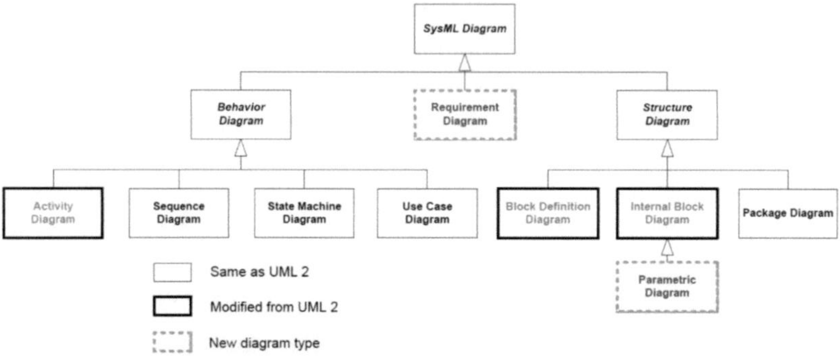

Figure 2-13: SysML Diagram Taxonomy [Omg18b]

Requirement Diagrams (REQ) are used to define the system requirements and their relationships with other model elements. *Block Definition Diagrams (BDD)* define the structure of the system by defining its components, their properties, operations, parts etc. *Internal Block Definition Diagrams (IBD)* represent the internal structure of the blocks and specify the connection between the blocks by using special interaction points in SysML called ports. *Parametric Diagrams (PAR)* contain the parametric constraints present in between different model elements. These constraints can typically consist of mathematical equations. *Package Diagrams (PKG)* help organise the model into packages. *Activity Diagrams (ACT)* are used to define different aspects of system behaviour in the form of control and data flows. These diagrams are useful to define the functional behaviour of a system or its elements. *Sequence Diagrams (SEQ)* define the interaction between different system components and belong to the behavioural description of a system. *State Machine Diagrams (STM)* consist of possible states a system can have along with possible transition events to trigger these states. This helps to build event-based behaviour of a system. *Use Case Diagrams (UC)* are used to define possible use situations of a system. In particular, they are used to define the context of the system and its functional requirement. They show the expectations of the different stakeholders of a system. Several pieces of research discussing the modelling process using SysML diagrams are mentioned towards the end of section 3.1 and the behaviour modelling approach for VR-models is presented in section 4.3.

2.5 Summary

The research conducted in this thesis touches multiple hardware, software and methodological aspects from different research fields. The purpose of this chapter was to familiarise the reader with fundamental models and terms that are referred to in the later part of this thesis. In this regard, this chapter has discussed the product development process models that are most relevant to this research. After the discussion of these models, a general introduction of VR technology is provided, its different hardware systems are discussed in detail, their benefits and shortcoming are briefly discussed as well. The method for the descriptions of VR-models developed in this thesis utilises MBSE and its implementation language i.e. SysML. Therefore, this chapter has also provided a brief introduction of SE and MBSE along with a brief discussion about its implementation. An overview of SysML and its diagrams is also provided. The reason for using SysML as implementation language in this thesis will be discussed in section 4.2.

This chapter has built the base for chapter 3 by defining the fundamental terms and models. The next chapter will take a deeper look at the current situation in the industry and the state of the art.

3 State of the art in product development

This chapter will provide an overview of the state of the art in the field of product development, the use of VR in product development and MBSE. The focus will be put on the topics that fit the scope of this thesis. The most relevant approaches and methods consisting of the use of VR and MBSE in the product development process are discussed.

3.1 MBSE in today's product development process

Systems Engineering (SE) focuses on an interdisciplinary documents based approach for the development of complex technical systems. Although MBSE builds on the developments in the field of SE, it focusses on modelling the system and use of models for performing different analysis, verification & validation activities etc. The modelling process may start with the definition of system requirements followed by the modelling of structural and behavioural aspects of the systems, different analyses up to verification and validation activities. The modelling starts at the conceptual design and continues throughout the development process. The developed models allow the design teams to achieve early validation, consistent systems interfaces, better communication between the developments teams from different disciplines and traceability throughout the development process. The goals of MBSE are improved communication among the development stakeholders, increased ability to manage system complexity, improved product quality, enhanced knowledge capture and reuse, increased productivity and reduction of risk [Inc15].

In light of the promises, benefits and potentials associated with SE & MBSE methods, there is an ever-increasing inclination of the industry towards their incorporation in the product development process. A study is conducted by [Gau15] to examine the recognition and establishment of SE in different industrial sectors as well as to identify the challenges associated with products of tomorrow. The study conducts interviews of key officials from various sectors i.e. mechanical and plant engineering, automotive industry, automation engineering, aerospace, equipment manufacturers and other vehicle manufacturing companies. During this study, the growing multidisciplinarity is identified as the most significant challenge in the development of the products of tomorrow. Analog to this challenge, the orchestration of multidisciplinary cooperation is identified as the most significant benefits of SE. This study, after evaluating the industrial feedback, deems SE as a necessary prerequisite for the development of complex and multidisciplinary technical systems of tomorrow.

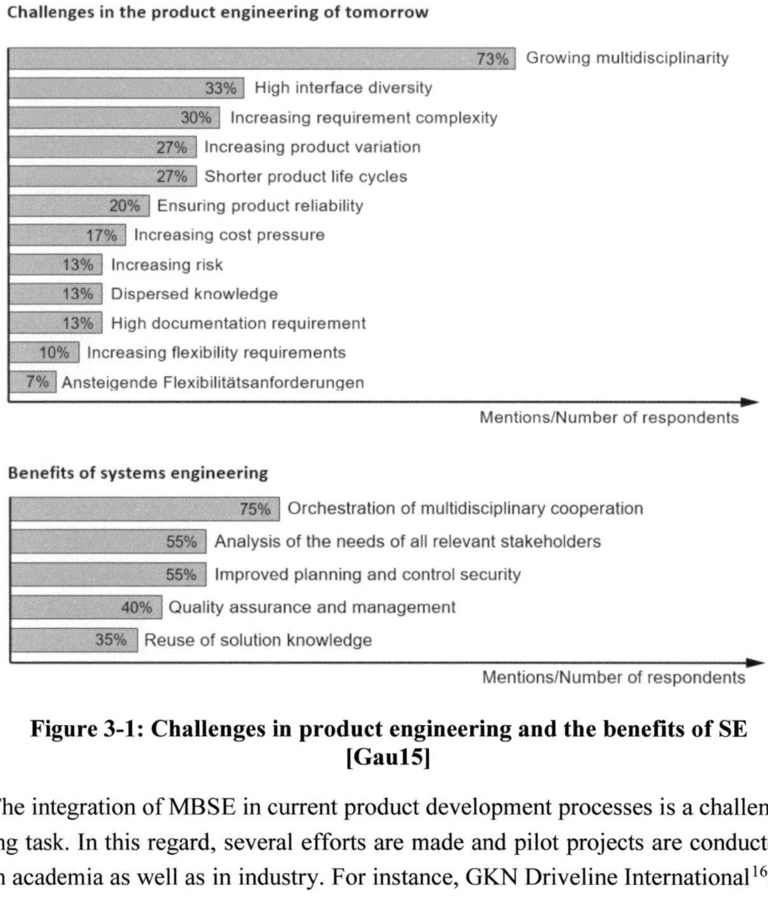

Figure 3-1: Challenges in product engineering and the benefits of SE [Gau15]

The integration of MBSE in current product development processes is a challenging task. In this regard, several efforts are made and pilot projects are conducted in academia as well as in industry. For instance, GKN Driveline International[16], a leading supplier of automotive driveline technologies, has started exploring the use of MBSE in the early development of its new intelligent systems [Ric17]. [Ric17] present the experience report about the adaptation of MBSE and simulation in the early stages of design. Starting with the modelling of system context, the functional structure and behaviour of the system are modelled. Logical solutions for sub-

[16] GKN: https://www.gknautomotive.com/ [last accessed on 09.03.2020]

functions are identified and combined to form a logical architecture of the product. This logical architecture serves as the base for further model development. Integration of the developed models with simulation tools like Matlab/Simulink helps to achieve the desired early simulations.

Daimler AG has been working on the SEED[17] project to realise a functional car prototype using MBSE [Epp18]. SEED uses SysML to model different views of the systems. The functional requirements of the customer are mapped to heterogeneous development functions of the system i.e. an automobile vehicle.

Miele[18] and Karl Mayer[19] are also working on the introduction of MBSE in their product development processes [Kle16]. As smart household appliances are expected to communicate with each other, this has led to increased complexity and moved Miele to incorporate interdisciplinary MBSE methods. Requirement management, incorporation of stakeholder requirements, system context analysis, modelling of logical structure and early simulations are achievable by using MBSE. Furthermore, the product quality can be improved and the complexity can be handled. For Karl Mayer, the early virtual validation of machine properties is a topic of interest that can be achieved through simulations. This is done by integrating different authoring, modelling and simulation tools with ModelCenter[20] to make parameter/information exchange between these tools possible. As a result, an early analysis of the design can be performed [Kle16].

To facilitate the incorporation of MBSE in the product development process, two recent collaboration projects (mecPro2 & FAS4M) between industry and academia have been carried out. MecPro2 (Model-based Engineering of Products and Production Systems) deals primarily with the development of cybertronics systems [Eig17a]. The goal of this project was to integrate the information and facilitate the data exchange throughout all the disciplines involved in the development process. This project uses a model-based approach for the description of the system

[17] SEED: Systems Engineering Enchantment @Daimler

[18] Miele & Cie. KG: A manufacturer of consumer appliances and commercial equipment

[19] Karl Mayer Textilmaschinenfabrik GmbH

[20] ModelCenter: https://www.phoenix-int.com/product/modelcenter-integrate/ [last accessed on 09.03.2020]

under development and proposes an architectural framework along with the general structure of its model framework (see figure 3-2).

Architecture framework of mecPro2 is a concept that contains the general instructions about defining the complete system starting from requirements to systems structure until the definition of detailed behaviour. This framework uses SysML as the description language for implementation purposes. *Profile & Ontology* contain domain-specific extensions, terminologies, concepts etc. to achieve consistency in the model and *View & Viewpoints* represent a certain aspect about an element in the system along with its relationships with other model elements.

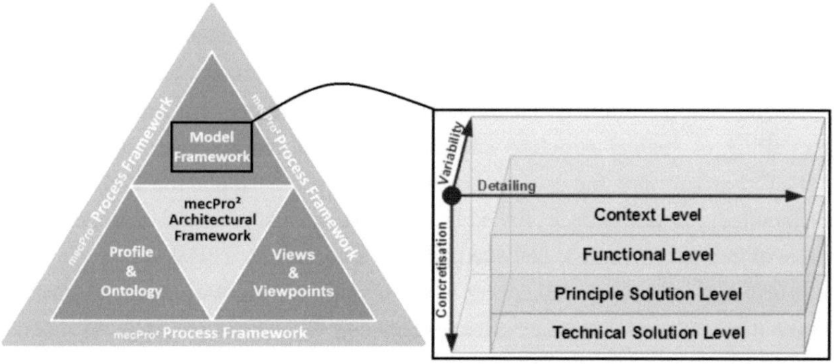

Figure 3-2: Architectural framework and it's model framework [Eig15]

Model Framework contains the backbone of the system under development and comprises of multiple layers of information (see figure 3-2 (right)). It contains four different levels of concretisations i.e. *Context Level*, *Functional Level*, *Principle Solution Level* and *Technical Solution Level*. In *Context Level*, systems requirements are extracted and converted to model-based requirements by considering different contexts of the system along its life cycle. *Functional Level* contains the solution functionality against a given context in the systems' life cycle and is kept solution independent. The technical aspects of the system that are necessary to realise the desired functions are considered and analysed in *Principle Solution Level*. Typically, there may exist multiple solution variants for a single function that can be analysed and evaluated to select the optimal solution. *Technical Solution Level* contains the technical components that realise a certain system function and delivers a system structure, where different system components are linked with each other over interfaces [Eig17b]. In this way, the architectural frame and model framework can be used to model a complete system. The use of SysML for modelling of these frameworks eventually established the feasibility of SysML for describing a complete system and the support for the development process.

In model-based development methods, one important aspect is the management of models that define the system. To manage these system definitions that are modelled in SysML, mecPro2 suggest a multilayer architecture for achieving a coupling between SysML authoring tool and the PLM system as shown in figure 3-3.

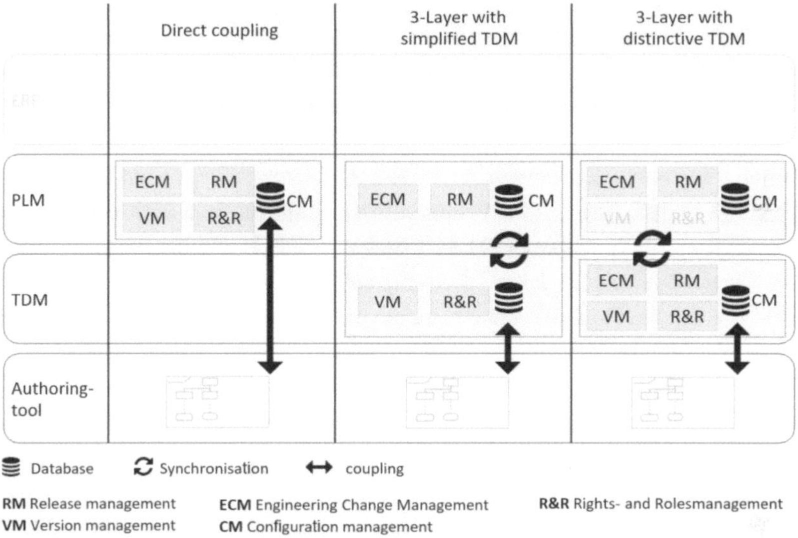

Figure 3-3: Multilayer architecture for the management of SysML models [Eig17b][Kir16][21]

The lowest layer contains the authoring tools used to construct the SysML models. *Team Data management (TDM)* builds the next layer that contains discipline-specific functions for the organisation of models and processes. PLM layer offers the possibility to integrate different disciplines through generic model management functions. As depicted in figure 3-3, the coupling can be achieved in multiple different ways depending upon the application and needed functionality. In *Direct coupling* scenario, the PLM layer handles all needed management functions to

[21] TDM is a locally available management system normally implemented by the tool vendor.

achieve the needed functionality. In the *3-Layer with simplified TDM* scenario, some of the data and model management functions are taken up by the TDM layer from the PLM layer. To have real-time updated data in different layers, a synchronisation between PLM and TDM is needed. Depending upon the coupling scenario, different management functionalities can be achieved. For instance, in scenario 3 (*3-layer with distinctive TDM*) a stronger TDM layer means that more of the management functions will be taken up by the TDM layer from the PLM layer. [Eig17b]

As the PLM systems today use a hierarchical structure for model management, a similar hierarchical structure in the SysML authoring tool can facilitate the information exchange between PLM and SysML authoring tool [Kir16]. Figure 3-4 shows an identical structure of PLM and SysML authoring tool that was evaluated inside a software demonstrator for PLM-SysML integration within the scope of mecPro2 project. The PLM structure in this example is from CIM DATABASE PLM[22] from the company CONTACT Software GmbH and the authoring tool is the Cameo Systems Modeller[23] from the company NoMagic. Such an identical structure can facilitate efficient integration of PLM system and SysML authoring tools by achieving bidirectional synchronisation of model elements and artefacts. Such integration can greatly facilitate the management of SysML models in well-established PLM systems as well as can enable a bidirectional exchange of information in real-time.

[22] CIM DATABASE PLM: https://www.contact-software.com/de/produkte/plm-software-product-lifecycle-management/ [last accessed on 30.03.2020]

[23] Cameo Systems Modeler: https://www.nomagic.com/products/cameo-systems-modeler [last accessed on 30.03.2020]

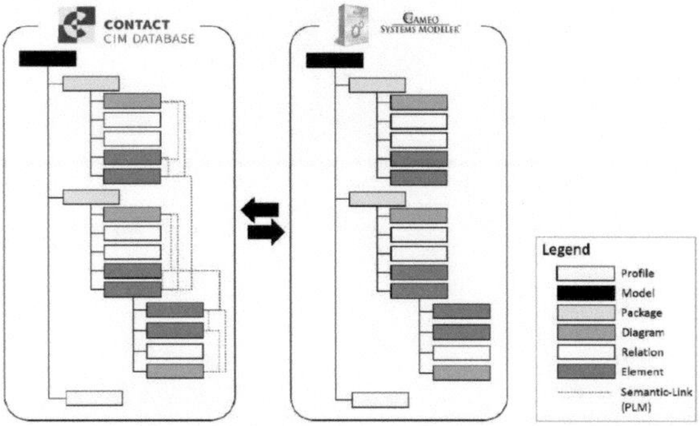

Figure 3-4: Identical structure in PLM software and SysML authoring tool [Kir16]

Furthermore, concerning the integration of PLM and SysML, there is a very important approach of System Lifecycle Management (SysLM) [Eig16]. This approach is based on an integrated, information-driven concept for achieving traceability over the complete system life cycle and to achieve higher performance of a product system (see sub-section 2.1.3 for details). This can be achieved by implementing a shared information core system by integrating different tools and data management systems used inside the product development process. The advantage of such a shared core system is the availability of real-time information in all different development teams about the status of development that eventually helps to avoid communication errors. The SysLM concept also helps to manage the complexity of the system, right from the start of development until the end of the product life cycle and ensures the availability of the right information at the right time of the development. The two-layer-model of SysLM can be seen in figure 3-5. The extended V Model for the development of multi-disciplinary products included in SysLM concept has already been discussed in sub-section 2.1.3.

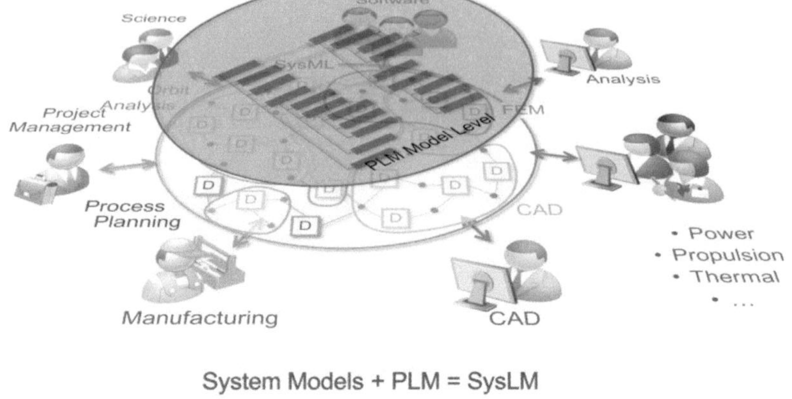

Figure 3-5: Two-Layer-Model of System Lifecycle Management [Eig16]

The fact that MBSE and the process of product modelling in MBSE are new to many of the mechanical engineers and product designers in the industry, even the wildly adopted modelling language in MBSE i.e. SysML is very abstract and generic for use by mechanical engineers. To bridge this gap, Functional Architectures of Systems for Mechanical Engineers (FAS4M) was a research project carried out recently by a collaboration consortium comprising of partners from academia and the industry [FAS20]. FAS4M identifies the need for bridging the gaps between MBSE models and the 3D-CAD-models, to facilitate the incorporation of MBSE in the development process. The direct transition from the product's abstract structure in the Block Definition Diagram (BDD) of SysML to concrete 3D-CAD models can lead to errors. Therefore, FAS4M works on the development of an intermediate solution in form of a modelling language i.e. MechML (Mechanics Modeling Language). MechML is built on the foundation of SysML, uses its core functions and contains model elements specific for the use of mechanical engineers. MechML also facilitates bidirectional information exchange between SysML modelling tool (Cameo Systems Modeler) and CAD software (CATIA V5[24]). This information exchange is carried out with the help of two plugins (one in SysML tool and second in CAD tool) as shown in figure 3-6.

[24] CATIA V5 : https://www.3ds.com/de/produkte-und-services/catia/produkte/v5/portfolio/ [last accessed 09.03.2020]

Figure 3-6: Bidirectional communication between SysML and CAD [Moe15b]

The FAS-method developed inside this project lays out a clear roadmap for the modelling of the functional architecture of a product [Dän14]. FAS-method also facilitates the use of free sketches in MBSE, so that incorporation of MBSE methods among mechanical engineers can be encouraged [Moe15a]. The software prototype implemented under FAS4M achieves the bidirectional communication between product shape in CAD and the functional architecture in MechML (extension of SysML). The results of the research conducted in this project are summarised in the form of methodological guidelines for the implementation of FAS-method and are documented under [Moe16].

MecPro2 and FAS4M have contributed to the integration of SysML models with PLM and CAD systems. Both of these projects build the base for the incorporation of SysML and MBSE methods in the current product development process. Besides these two projects, there are several contributions available that talk about the use of modelling and MBSE methods in product development. [Fol10] discuss the use of SysML for the development of mechatronic systems by modelling an example product i.e. washing machine using SysML diagrams. The different diagrams of SysML are used in modelling and their model elements are briefly discussed. This approach presents a discipline neutral modelling approach and identifies SysML as a suitable language to model complex mechatronics systems. A more detailed overview of the SysML behavioural diagrams and the behavioural modelling process is presented in [Sil11]. In this work, the use of behavioural diagrams and the modelling process are elaborated using an example model of a subpart of an audio player. Although both of the last-mentioned works describe the modelling process in SysML, they do not provide any way for the simulation of these models, instead, they identify the simulation as a topic for further research. SysML was developed as a graphical description language and its models were not supposed to be simulated. However, seeing the potentials of SysML and the interest of industry towards it, there have been multiple efforts to make SysML models executable. [Ban16] achieve coupling of SysML modelling tool and Matlab/Simulink to simulate the SysML models. In this work, the instructions based communication (carried out by UDP[25] driver) between SysML modelling tool and Matlab is achieved. Discrete-event simulation of SysML models can be achieved by transforming these models into the Arena[26] models [Bat12]. Such a transformation can be performed using the Atlas Transformation Language [Alt20]. Once the SysML models are transformed into Arena, they can be simulated inside Arena and a discrete-event simulation can be obtained. [Bül14] simulate SysML models by transforming them into MODELICA[27] (an executable language). There are more contributions [Gro15][Sin11][Fou12][Sch09] that try to simulate the SysML models either through model transformation or through an integration with an executable language/tool.

To use MBSE models for achieving a simulation in VR, MASCARET[28] framework is developed. The virtual environment is modelled using graphical diagrams of UML and these diagrams are converted into a simulator model. The conversion of the models into the simulator model is done by using standard XMI (XML Metadata Interchange) format [Che12]. [Abi15] extend MASCARET framework and develop a method to achieve a simulation in VR (see figure 3-7). This simulation model is first developed in SysML and later transformed into the Arena simulation model.

[25] UDP: User Datagram Protocol
[26] Arena Simulation Software: https://www.arenasimulation.com/ [last accessed 09.03.2020]
[27] Modelica: https://www.modelica.org/ [last accessed 09.03.2020]
[28] MASCARET: A pedagogical multi-agents system for virtual environment for training [Buc04]

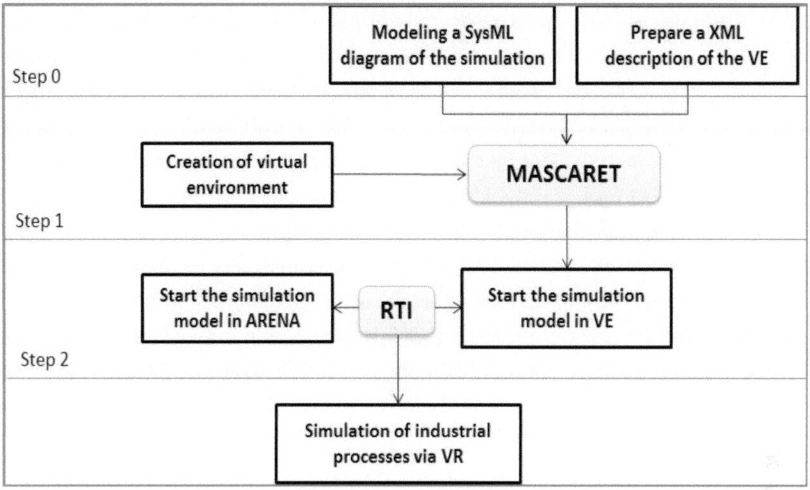

Figure 3-7: Main steps of the methodology [Abi15]

In the continuation of [Abi15], [Abi16] use the same method to achieve a simulation of production lines in a virtual environment. The SysML model describing the VR simulation is transformed into the Arena model. The coupling of the Arena model with the virtual environment makes it possible to achieve discrete-event based simulation of production lines in VR.

Concerning the increasing adoptions of MBSE models and methods in industry, an MBSE based method for simulation of VR-models can greatly facilitate the incorporation of VR in industry. The available methods and modelling tools for implementing MBSE have attained enough maturity to facilitate direct simulations in VR. After discussing the state of the art about MBSE, the rest of this chapter will discuss current applications and general practices regarding the use of VR in product development.

3.2 Virtual Reality in product development

The early identification of the faults and discrepancies in the design is vital to the product development process because the later identification of a fault can be very expensive [Ste04] or sometimes difficult/impossible to correct. This has led to extensive use of digital models and methods for supporting the product development process. Along with these models, new technologies are incorporated to facilitate early evaluations. Among these technologies, Virtual Reality (VR) and Augmented Reality (AR) technologies are increasingly finding their place in the product development process. Both these technologies work with the digitals models available during product development and offer various possibilities to examine and analyse the design. The discussion by [Rad14] about VR identifies it as an important support tool in virtual product development. However, the preparation of VR-models for use in product development is a difficult task. The current solutions available for integration of VR into the current product development process are not satisfactory and also not suitable for use by product designers [Sta09a] and [Sta09b]. [Sta09b] suggest the use of grid technology to exploit the potentials of virtual product creation and develop the ProGRID. One interesting aspect of ProGRID that is very relevant to this thesis is the realization of simulation data management (SDM) system.

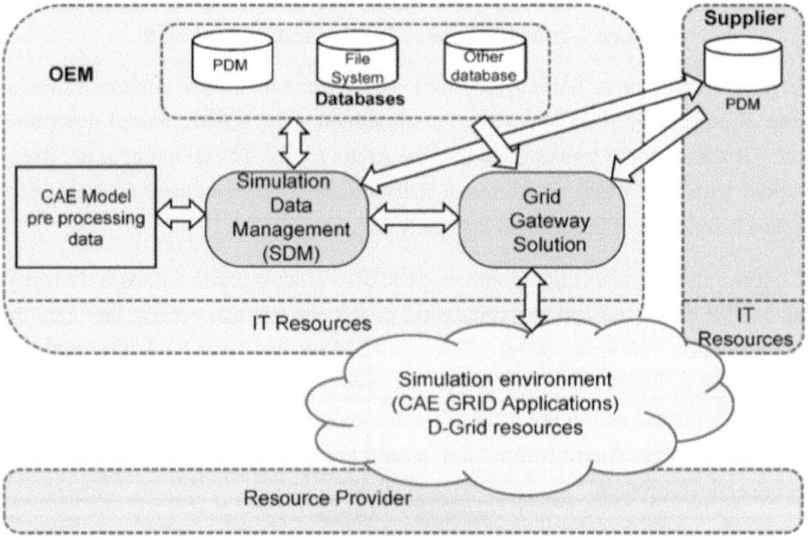

Figure 3-8: Simulation data management system with access to grid resources [Sta09b]

The overall data about the system is stored inside common databases and at the time of simulation, the necessary data is collected. In this way, common databases can be obtained that facilitate the availability of always up-to-date simulation related data at the right time to the right stakeholder. This can also help to avoid miscommunication between development teams.

Usually, 2D computer screens are used to visualise simulation data or product models. The visualisation of a 3D geometrical model of a product on a 2D computer screen can at times provide a misleading understanding of the product. VR enables interactive visualisation of 3D models by allowing an interactive 3D visualisation [Elc11]. Such visualisation can provide the designers with useful information about the design e.g. the actual dimensions of the product, facilitate collaborate design reviews etc. Furthermore, the incorporation of product behaviour inside VR can provide a real-time alternative to a physical prototype. In the last decade, VR has been used extensively during the product development process in different application areas such as:

- design evaluation [Ye07] [Par08] ,
- virtual prototyping [Zor03],
- production planning [Sch05] [Ter15],
- assembly planning [Bul00],
- virtual assembly [Jay99] [Set11],
- manufacturing [Cho15],
- ergonomic evaluations [Whi04]
- learning & education [Gor17] [Abu11] [Ver17],
- acoustical product simulations [Hus14] [Sie16],
- . . . and many more.

Furthermore, in light of the life-phases of a product defined in VD2221 (see figure 2-1), potential application domains of VR can be extracted as shown in figure 3-9. In product development, VR can be used to achieve different simulations and also for performing visual inspections. Manufacturing and assembling processes can be visualised in VR as well as the flow of materials and logistics information can also be simulated. In marketing and sales, VR can offer a cost-effective solution for remote presentation of the product to users or stakeholders. The operation of the product can also be evaluated in VR by performing ergonomic evaluations, safety tests, training and simulations of the maintenance & repair procedures. Finally, the disassembly and recycling possibilities of a product can as well be investigated in VR which are important topics considering the sustainability of the design.

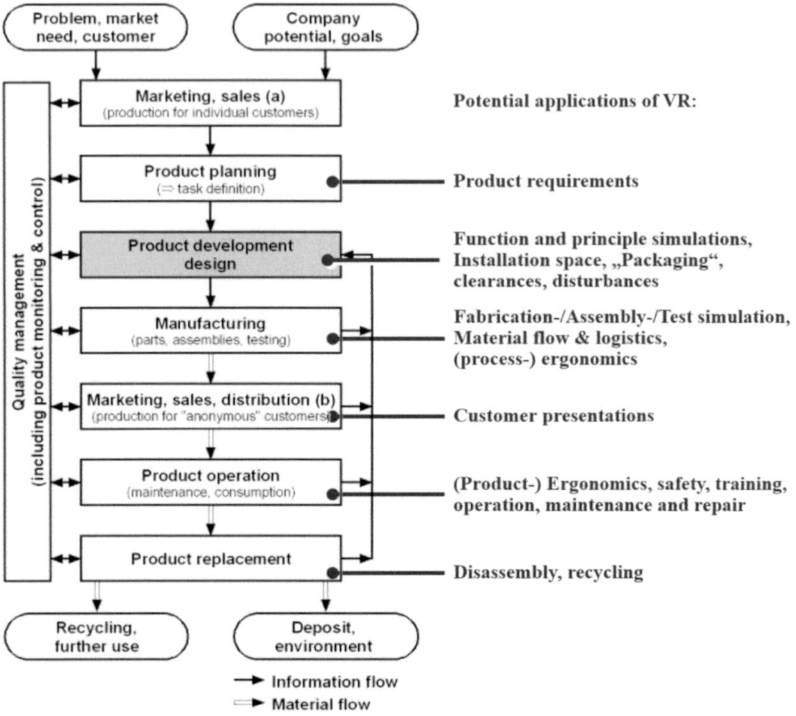

Figure 3-9: Potential applications of VR (extended from figure 2-1)

Based on the discussed applications in this section, it is clear that VR has penetrated in most of the disciplines and finds its application throughout the product life cycle. The next section will address the conventional process for creating VR-models and an overview of the current VR-model preparation practices will be provided.

3.3 Methods for the preparation of a VR-model

Typically, the construction of a VR-model contains two important parts. The first part deals with the visual representation of the geometrical objects and the second part deals with the construction of behaviour of the objects or the addition of simulation/animations. Other than these two parts, there may exist further parts e.g. haptic feedback devices etc. but are not discussed here. To import a product model that is (usually) available in the form of CAD models, the VR-software offers the possibility to import these models. However, usually, the models cannot be imported in native CAD formats, instead, the geometric exchange formats like STEP, VRML, JT etc. are used. Different geometric exchange formats can carry different details and levels of information about the original CAD model. An overview of the 3D data exchange formats relevant to CAD to VR transfer can be seen in table 1.

Table 1: 3D exchange formats relevant to CAD-VR conversion [Lor16]

	JT	STEP	IGES	VRML
Geometry	Yes	Yes	Yes	Yes
Materials	Yes	Yes	Yes	Yes
Textures	Yes	Not implemented in CAD exporters	No	Yes
Animations	No	Not implemented in CAD exporters	No	Yes
Constraints	No	Not implemented in CAD exporters	No	Limited
Kinematic relationships	No	Not implemented in CAD exporters	No	Limited

The choice of the data exchange format for a given application not only depends on the information carried by that format but also on the VR-software and CAD tool at hand. Different CAD tools offer different export formats and similarly, different VR-software can import different formats (see figure 3-10). An overview of the renowned CAD and VR tools along with their export and import possibilities is provided by [Mah16] and can be seen in figure 3-10.

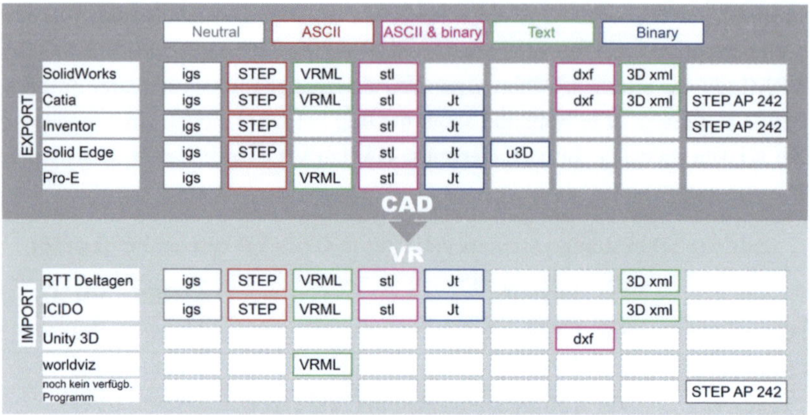

Figure 3-10: Overview of renowned CAD and VR tools along with the import and export possibilities [Mah16]

An overview of a few VR-software that are currently available in the market and their support for different VR-systems can be seen in table 2.

Table 2: Support of VR-software (rendering tools) for different VR-systems

VR-software	Type	CAVE	Powerwall	HMD	sVR
RTT Deltagen	Commercial	Yes	Yes	Yes	No
IC.IDO	Commercial	Yes	Yes	Yes	No
Visionary Render & Visionary Render 2	Commercial	Yes	Yes	Yes	?
VR Studio	Commercial	Yes	Yes	?	?
Vizard	Commercial	Yes	Yes	Yes	No
Unity3d	Commercial / Free*	No	No	Yes	Yes
Unreal Engine	Commercial / Free*	No	No	Yes	Yes
Free*: for educational, personnel and companies with a limited annual turnover					

The information presented in this table is referenced from the tool vendor websites (last accessed on 09.03.2020). RTT Deltagen[29], IC.IDO[30], Visionary Render[31], VR Studio[32], Vizard[33] are available commercially and mainly focus on the CAVE and Powerwall based VR-systems. Unity3d and Unreal Engine are game engines that are available both commercially and as free versions (for academic use). Both of these game engines do not offer direct support for CAVE or Powerwall type VR-systems and instead, focus on smartphone and HMD based applications development or computer games. It is evident from table 2 that there is no VR-software currently available that can cover all the available VR-systems.

Furthermore, not all of the VR-software can import all 3D data exchange formats used in CAD tools. At times, an intermediate conversion tool is needed. For instance, 3D content creation software like Blender[34] (open source) or Autodesk Maya[35] can import several 3D data exchange formats and act as an intermediate tool between CAD and VR. After importing the CAD geometries using exchange formats, both of these tools facilitate manipulation, simplification and export of the geometrics in such an exchange format that can be directly imported into VR-software. A more detailed overview of the geometric exchange formats is provided by [Mch08] that enlists the standards, types, contained information and owning tool of these formats.

Based on the information presented in table 1, table 2 and figure 3-10, it can be concluded that the choice of VR-software depends on multiple factors i.e. the CAD tool used inside the company, supported exchange formats, supported VR-system type etc. Furthermore, the geometrical representation of a VR-model designed for one particular VR-system may not be reusable in other VR-systems due to the poor interoperability between different VR-software.

---

[29] 3DEXCITE DELTAGEN: https://support.3dexcite.com [last accessed on 09.03.2020]
[30] IC.IDO: https://virtualreality.esi-group.com/head-mounted-displays-virtual-engineering [last accessed on 09.03.2020]
[31] Visionary Render 2: https://www.virtalis.com/virtalis-visionary-render-2-visualisation-industry-4-0/ [last accessed on 09.03.2020]
[32] VR Studio: https://epiito.com/vr-studio [last accessed on 09.03.2020]
[33] Vizard: https://www.worldviz.com/virtual-reality-software-features [last accessed on 09.03.2020]
[34] Blender: https://www.blender.org/ [last accessed 09.03.2020]
[35] Autodesk Maya: https://www.autodesk.de/products/maya/overview [last accessed 09.03.2020]

The visual model in VR mainly consist of geometrical models and the simulations/animations are added separately. After the construction of the geometrical model in VR, the next step is to add behavioural representations for the given model. This can be achieved by adding the animations/ simulations manually or by programming or by adding executable scripts. Table 3 is based on a literature review and provides an overview of past methods developed for achieving behavioural simulation inside VR. The researches published in the past twenty years that talk not only about the visual representations of products in VR but also build the behavioural representations are considered. Table 3 does not list down all the methods developed to date for the behavioural descriptions of VR, instead presents a general idea about the current practices. The focus is put upon extracting the goal of the research, VR-software used, the type of VR-system used, the method used for achieving simulation as well as the simulation tool used. Furthermore, the contents of each research relative to the product, actor and environment are listed. It is also checked if the conducted researches perform any division of the complete VR-model into smaller isolated reusable modules / sub-models. To show the information originating from nineteen research articles on a single page, some abbreviations and symbols are used. These abbreviations and symbols are:

- VADE – A Virtual Assembly Design Environment [Jay99]
- CASUS – Computer Animation of Simulation Traces [Lus97]
- em-Plant – new name *Plant Simulation*[36] by SimPlan AG
- VDTC – Virtual Development and Training Centre (ELBEDOME[37])
- Vizard – Vizard VR by wordviz[38]
- Virtools – *Virtools* by Dassault Systems
- TCLV – Teamcenter lifecycle visualization 9.1 by Siemens
- Revit[39] – Building Information Modeling Software by Autodesk
- CACE – Computer-Aided Control Engineering
- E, T & C – Environment, Technical & Cultural
- P, A & E – Product, actor and environment

[36] Plant Simulation: https://www.plant-simulation.de/ [last accessed 09.03.2020]
[37] ELBEDOME: https://www.elbedome.de/ [last accessed 09.03.2020]
[38] Vizard: https://www.worldviz.com/vizard-virtual-reality-software [last accessed 09.03.2020]
[39] Revit: https://www.autodesk.de/products/revit/overview [last accessed 09.03.2020]

- In – Simulation model developed and executed inside the VR-software
- Out – Simulation model developed and executed outside the VR-software
- * – Asterisk sign is added if the research does not state the used technology, but it can be extracted by the reader

The researches mentioned in table 3 by large cover different aspects related to product development e.g. assembly planning, product evaluation, ergonomic evaluation, decision making support, acoustical evaluation, design reviews etc. The use of VR for training and educational proposes is also included in this review.

Table 3: Conventional ways of describing VR applications

Goal	VR-software	Sim. method	Contents			Division	VR-system	Source
Virtual assembly planning	VADE	In	P	--	--	--	HMD	[Jay99]
Virtual factory for manufacturing	CASUS	In	P	--	E	--	Powerwall*	[Mue02]
Virtual factory & assembly simulation	--	Out / eM-Plant	P	A	E	--	--	[Sch05]
Educational and entertainment	--	--	P	--	--	✓	--	[Wal06]
Product design evaluation	OpenGL and GLUT	Out/ C, C++	P	--	--	--	--	[Par08]
Embedded systems evaluation in VR	VDTC	Out	P	--	--	--	2D Display	[Köp09]
Product acoustical evaluation	RTT Deltagen	In	P	--	--	--	CAVE	[Hus09]
Ergonomic evaluation	--	--	P	A	--	✓	HMD	[Bor09]
Learning and training in eng. edu.	Autodesk Showcase	--	P	--	--	--	Powerwall*	[Abu11]
Simulation of automated workspaces	Vizard	--	P	--	E	--	HMD	[Nov11]
Product interface evaluation	Virtools Dev	Out / Matlab/ Simulink &	P	--	--	--	AR glasses	[Bru13]
Product acoustical evaluation	RTT Deltagen	In	P	--	E	--	CAVE	[Sie16]
Early design decision making	TCLV	--	P	--	E	✓	CAVE	[Ber17]
Learning and training in engineering	Unity, Unreal	In	P	--	E	--	HMD, sVR	[Ver17]
User experience evaluation	Vizard	In	P	--	--	--	CAVE	[Son18]
Digital factory for training and maint.	Unreal Engine 4	In	P	A	E	--	HMD	[Sha19]
Social interaction in VR	Unity	In	--	--	--	--	HMD, sVR	[Hol19]
Service design prototyping	Unity	In	P	--	E	--	HMD	[Nam19]
Analysis of VR workflow	Revit, Unity, Unreal	In	--	--	--	--	HMD, sVR	[Pra18]

A broad spectrum of VR-software is used that are either commercially available or were developed as a result of conducted research. As far as the method of achieving simulation in VR is concerned, there can be an easy differentiation made between them i.e. the simulations/animations are either directly added inside the VR-software or performed external to the VR-software. The simulation inside VR-software is achieved by using programming, visual scripting or flow charts depending upon the possibility offered by VR-software at hand. The out of the VR-software simulation is achieved by coupling a simulator with VR-software. If the simulation is done outside VR-software in any given study, then the simulation tool is also mentioned inside the table. Most of the mentioned works treat the complete VR-model as one entity except for [Wal06],[Bor09] and [Ber17]. [Wal06] divide the virtual objects into the environment, technical and cultural groups before saving them inside a self-made database. [Bor09] perform an ergonomic evaluation of a product in the presence of a human actor. [Ber17] perform product design review of a pump as a product inside the assembly line as its environment. In general, the discussion about the reusability of the developed applications is limited. [Mue02] and [Wal06] claim to achieve reusability of VR-model contents by developing model database and libraries. However, the reuse of the created content across different VR-systems is not discussed explicitly. [Bru13] add simulations in VR by integrating product functional behaviour models from Matlab/Simulink inside a mixed reality setup using AR glasses. This method claims to be reusable in different VR-systems but no validation is provided. A limitation of this method is that the simulator model has to be developed separately if the product developer is not using Matlab/Simulink. The *Contents* column in the table indicates that the product is the subject of the research primarily and the examination of actor & environment is limited. Based on the brief comparison in table 3, it can be concluded that

- the choice of VR-software is highly application and VR-system dependent,
- the simulation is achieved by using different methods and there is no general method available,
- product is mainly the focus of the examination and limited examination of actor & environment is included,
- there is no uniform method for division of the complete VR-model and
- reusability of the VR content is not demonstrated explicitly.

Although some of the commercially available VR-software offers built-in physics calculation modules, they are mainly limited to physical phenomena like collision detection and rigid body dynamics [Bru10]. Furthermore, the use of these physics modules are subjected to the inside VR-software development based on programming, visual scripting etc. that is again not suitable for reuse in different VR-systems. Therefore, the idea of an external simulator integration with VR is appealing concerning the physical calculations as well. Integration of external simulator can facilitate the product designer to use the simulation models already available during product development as well as can facilitate the reusability of developed simulation models in different VR-systems.

It is evident from the discussion in this section that there is no generic method available for the preparation of VR-models. Previous researches use different VR-software, simulators and exchange formats to prepare VR-models. As indicated in table 2, one VR-software cannot support all available VR-systems and thus, interoperability in different VR-systems cannot be achieved. Therefore, a new generic method for the preparation of a VR-model is needed that is independent of VR-software and VR technology and which can reduce the preparation effort by achieving reusability in one VR-system and across different VR-systems. Possibilities to reduce the preparation effort for VR-model are the use of the simulation models that are already available during product development and the reuse of the contents of already created VR-model. The contents of a VR-models can be reused on two different hierarchical levels as indicated as *Level 1* and *Level 2* in figure 3-11.

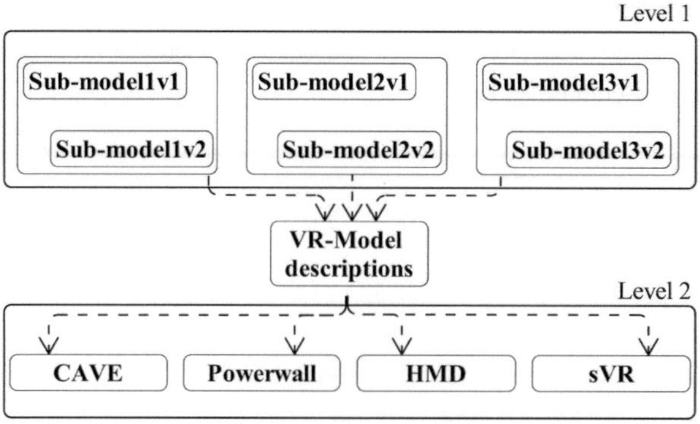

Figure 3-11: Hierarchical level of reusability of VR-Model

The first level includes a detailed view of the VR-model, where the complete model is divided into small isolated reusable modules/sub-models. These isolated sub-models can be reused to form different use cases while being in the same VR-system. The second level addresses the reuse of VR-model descriptions in different VR-Systems irrespective of the VR-software used.

So far an overview of the state of the art in the field of VR in product development is provided in this chapter with the conclusion that there is a need for a new generic method for the preparation of VR-models.

3.4 Summary

This chapter has provided an overview of the current developments in the field of MBSE and its integration into the product development process. The potentials of handling the complexity of technical products and of achieving traceability throughout the development process have made MBSE an important topic of research in academia as well as in industry. The industrial companies have already started carrying out pilot projects to examine the effectivity of MBSE in the development process and efforts for the incorporation of MBSE methods are on full swing. Furthermore, preliminary studies have elaborated the behaviour modelling process in SysML and the use of SysML & MBSE methods to drive a simulation model in VR is also put to test. However, the currently available methods use different transformations between SysML models and VR. Some of the information about the product model can be lost during such a transformation process. Furthermore, the physical calculations cannot be directly incorporated in SysML which also remains an open topic for further research.

Although the use of VR in product development is very promising, the preparation of the VR-model remains a challenge. This is due to the poor standardization of VR-software and the resulting interoperability issues between them. The construction alone of the geometrical model in VR makes the choice of VR-software a challenging task because of factors like CAD tool, supported exchange formats, VR-system type, application area etc. The behavioural model in VR relies most of the time on programming or scripting based solutions that can offer very limited reusability. Thus, every new use case /scenario of a product in VR requires

- remodelling,
- reprogramming and

- sometimes even reconstruction of the complete VR-model.

This situation at hand establishes the need for a new generic method that can

- reduce the overall VR-model preparation effort,
- support all available VR-systems,
- make VR content reusable in different VR-systems regardless of the used VR-software and
- can facilitate the incorporation of physical calculations on the geometric objects.

Based on these requirements, the relevant research questions are formulated and presented in the next section.

3.5 Research questions

This research focuses on the development of a method for the description of VR-models to facilitate the use of VR in product development by reducing the needed effort for preparing the VR-models. The preparation effort can be reduced considerably by reusing the simulation models available during development and by reusing the already created VR contents. This reusability has two levels of hierarchy i.e. on VR-system level and within one VR-system on the sub-modules/sub-models level (see figure 3-11). Furthermore, the new method should also include the possibility of performing physical calculations on the geometrical objects in VR with ease. To fulfil the requirements of the new method, the following research questions should be addressed during this research.

1. What can be a useful division of a VR-model[40] to make elements (sub-models) of the complete model reusable? *[addressed by chapter 4]*
2. How to create the separate isolated descriptions of sub-models including behavioural descriptions? *[addressed by chapter 4]*
3. How to achieve a generic (formal) description of a VR-model including (physical) behaviour description for use in product development that can be reused as well? *[addressed by chapter 5 and 6]*

To address the above-mentioned research question, there are further sub-questions that should be addressed i.e.

I. How can the generic interactions between the sub-models be modelled in case the complete model consists of sub-models? *[addressed by chapter 4, 5 and 6]*

II. How can the physical behaviour of sub-models be incorporated inside VR – which needs real-time conditions? *[addressed by chapter 5]*

III. How can these behaviour descriptions be used as the core of the simulation process and cooperate with VR-software in different VR-systems (CAVE,

[40] Also see definition in section Abbreviations and terms

HMD, Powerwall and sVR) to achieve an interactive simulation in VR? *[addressed by chapter 5 and 6]*

In the following part of this thesis, these research questions will be addressed and discussed one by one.

4 Method

This chapter presents the method developed for the descriptions of VR-models and attempts to address the research questions 1, 2 and the sub-question I defined in section 3.5. In light of the state of the art in section 3.3 followed by the discussion, a new method for the description of VR-models is necessary. The new method should address the current challenges regarding the preparation of VR-models as well as the research questions mentioned in section 3.5. Therefore, the new method shall

1. enable reuse of VR content within one VR-system,
2. enable reuse of VR content across different VR-systems,
3. reduce overall VR-model preparation effort and
4. offer easy integration in the product development process.

The requirements numbered 1 and 3 are addressed in this chapter whereas 2 and 4 will be discussed in chapter 5 & 6. The first section of this chapter talks about the division of the complete VR-model into reusable modules, the basic model for the construction of product use cases, the possible outcome of such use cases and the aspect of reusability of VR contents within one VR-system. The second section provides a general idea for the preparation of a VR-model while using SysML models. Section three develops the modelling methodology to describe the structure and behaviour of sub-models that resulted from the VR-model division and also includes the definition of interfaces between these sub-models. Furthermore, it contains a comparison of two different behaviour model execution architectures, as well as the automatic parallel execution and generation of use cases of product in VR. The fourth section summarises the findings and will note down the general guidelines for the implementation of the presented modelling approach.

4.1 Division of complete VR-model

It has already been discussed in chapter 1 that the multidisciplinary and complex products of today can no longer be treated in isolation. To develop a successful product, it is important to consider the context of a product consisting of the life-phase specific actor(s) and the environment. A basic evaluation model in the next sub-section discusses the expectations and possible outcome of a product evaluation use case that contains these three aforementioned sub-models. This idea can be extended far beyond physical products and their immediate environments, e.g. by covering Product-Service Systems (PSS) in a VR simulation [Exn19].

4.1.1 Basic model

In any given life-phase of a product, life-phase specific actor(s) and environment make the context of a product. The effectivity of contextual use cases for the evaluation process can be understood employing the basic model in figure 4-1 that shows two different use cases of a product, i.e. using the product and service it, along with respective context.

- *E* represents all inputs to the system – intended as well as unintended – while *A* represents all outputs produced by the system [Mah17a]
- The context can be understood as a use case specific environment along with the respective actor(s), e.g. a fabrication machine operator, an end-user or a service technician. The context also contains the environment that is not necessary for functional behaviour but has to be considered concerning the disturbances [Mah17a]

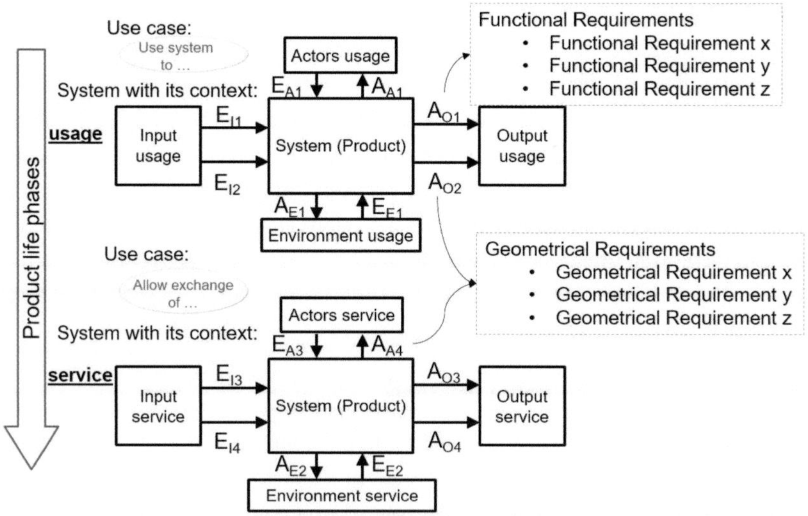

Figure 4-1: Product context in different product life-phases [Mah17a]

In each of these two use cases, the product can have one or more actors that interact with it. Similarly, it can find itself in a certain environment. The most critical thing to extract from these two use cases is the changing product requirements. For instance, in the usage phase of the product, the designer focuses on the functionality of the product and tries to address the needs of the user. Mainly the functional requirements i.e. efficiency, ergonomics, aesthetics etc. of the product are in focus here. This can change in the service phase of the product, as the product comes in contact with a technician in this phase. The designer will here have to address the needs of the technician e.g. easy diagnosis capabilities, easy access to out of order components and their easy replacement, etc. This may bring the geometrical requirements of the product in focus. Similarly, a product in its every life-phase can have different use cases and in all these use case situations, a different context of the product has to be considered. Figure 4-1 depicts the formal representation of the relational properties introduced by [And96] that were further developed by [Web07]. Building similar use cases like the ones discussed here can help the designer verify product requirements or maybe discover new requirements that are not yet considered in the design.

4.1.2 Division of VR-models

As discussed in the last sub-section, the three most important components of a product evaluation use case are the product itself, its environment and actor(s). In different use cases over the life span of the product, the actors can change as well as the environment. Furthermore, the product may itself consist of multiple design versions that have to be evaluated. Hence, the substitution and reuse of these three components are vital for building dynamic use cases for the sake of product evaluation in VR. To achieve such a dynamic configuration of different product evaluation use cases, one possible solution is to divide the complete VR-model into smaller reusable modules or sub-models. Therefore, the core of the new methodology lies in the fact that the complete VR-model is divided into sub-models of product, actor and environment (the idea was first presented in [Web16]). These three models are referred to as *sub-models* in the rest of this thesis. The main motivations for performing such a division are the following:

- The fact that the product will be continuously developed and at any given stage in development, if there is a use case of product needed in VR, it should be possible to substitute the existing product model inside the complete VR-model with the new one without having to make major adjustments to the overall model

- One way to reduce the preparation effort of VR-model is by achieving the reusability of the complete VR-model or parts of it. This division can allow us to deal with the sub-models individually and can thus enable the substitution/reusability of these sub-models

In light of this division, the external conditions mentioned in the CPM/PDD approach in sub-section 2.1.4 are revisited and further extended as presented in [Lie17]. The original approach considers the systems lying outside the boundary of a product as neighbouring systems and identifies their influence on product properties as external conditions. The originally defined external conditions are further extended to incorporate the actor(s) and the environment to include them as the neighbouring systems of the product. Figure 4-2 shows the extended external conditions that identify the actor and the environment as two important neighbouring systems of a product.

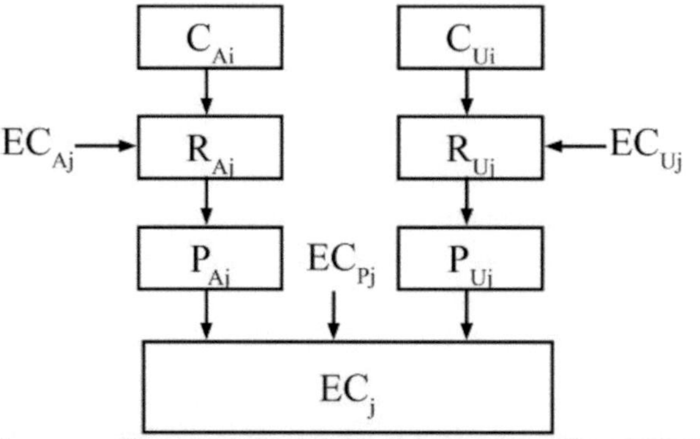

C_{Ai}, C_{Ui}: Characteristics (Merkmale) for actor and environment
P_{Aj}, P_{Uj}: Properties (Eigenschaften/is-properties) for actor and environment
R_{Aj}, R_{Uj}: Relation between characteristics and properties for actor and environment
EC_{Aj}, EC_{Uj}: External conditions for actor and environment
EC_{Pj}: External conditions for the product, without direct consideration of actor and environment
EC_j: External conditions with consideration of product context

Figure 4-2: Extended external conditions [Lie17]

The extended external conditions include the actor and environment model that allow the validation of product properties with the interplay of neighbouring systems. Similar to the product, actor and environment may contain their own external conditions as well as characteristics and properties. The separate representation of actor and environment in external conditions builds the base for a dynamic VR-model that can allow the contextual evaluation of product properties. However, caution must be exercised in the understanding of new external conditions i.e. the explicit definition of the impact of actor and environment in no way can cover all the external influences on a product. There may well be more influences originating from other sources which are mentioned as $\mathbf{EC_{Pj}}$ in figure 4-2. Figure 4-3 is the extension of the product development process schematic from [Web05] and contains the new extended external conditions.

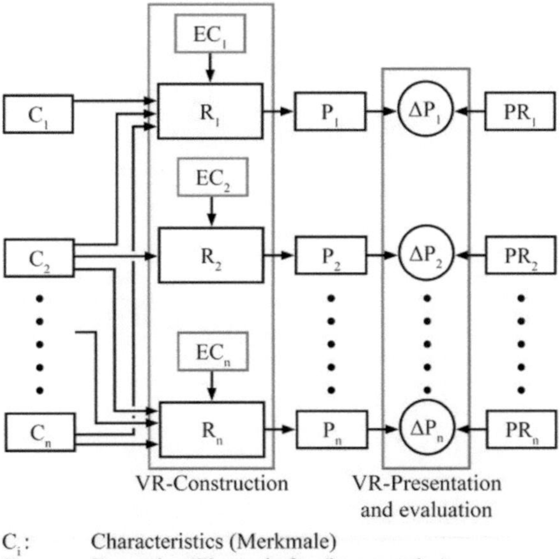

C_i: Characteristics (Merkmale)
P_j: Properties (Eigenschaften/is-properties)
R_j: Relation between characteristics and properties
EC_j: External conditions
ΔP_j: Difference between is and required properties as the driver of development process
PR_j: Required properties

Figure 4-3: Overall extension of CP-model of [Web05] as presented in [Lie17]

The new external conditions ($\mathbf{EC_j}$) here are the ones presented in figure 4-2. $\mathbf{PR_j}$ are the required properties that can be understood as an analogy to the requirements from the product. $\mathbf{P_j}$ in the above figure are the calculated properties or as-is-properties against the given characteristics under the consideration of extended external conditions. [Web11] suggest the use of VR as a tool for the virtual development process and its representation contain VR specific aspects that can be seen in figure 4-4.

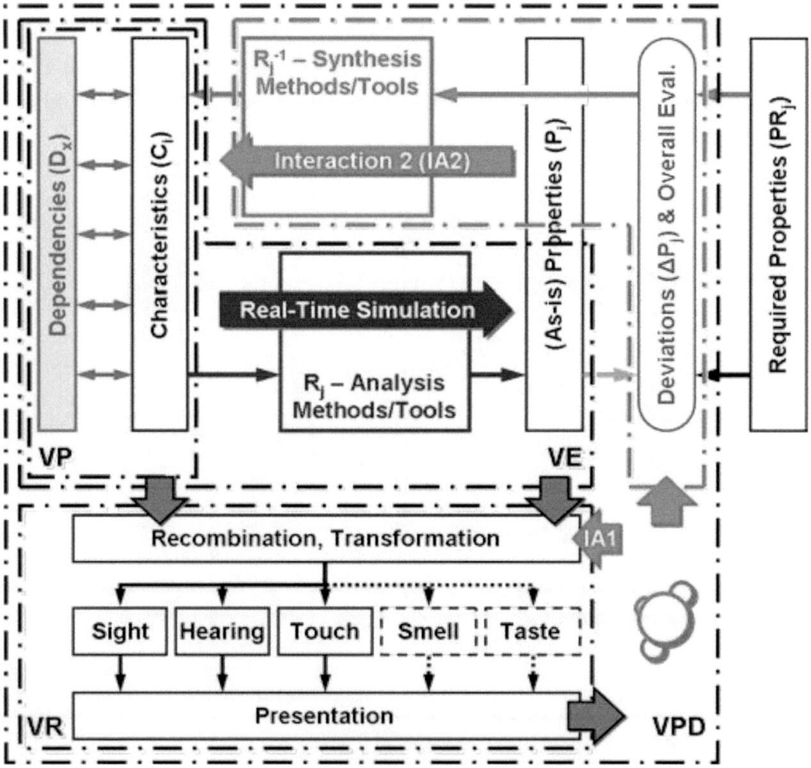

Figure 4-4: Virtual Reality (VR) as a tool of Virtual Product Development [Web11]

The difference *(ΔP_j)* between is-properties and the required properties is the driver for the development process. This difference can be a visual result of a product evaluation performed by using a VR-model. The original CPM/PDD approach is not limited to any particular method for calculating the properties from characteristics, however, in VR this is conducted on the base of visual presentation. Therefore, any shortcoming in design or product interactions can be seen visually (*Deviations (ΔP_j) & Overall Eval.* in figure 4-4 or *VR presentation and evaluation* (ΔP_j) in figure 4-3) in VR and it drives the development process. The VR-model division together with the CPP/PDD approach with extended external conditions build the base for product evaluation in VR and is the methodological backbone of this thesis. Another way to look at the benefits of performing such a division and using sub-models to form different use cases in VR can be understood graphically as shown in figure 4-5. Considering the left side of this figure, a scenario of a product's contextual evaluation in VR consists of actor, environment and system (i.e. product). Construction of this scenario in VR can help to validate product requirements associated with this particular scenario.

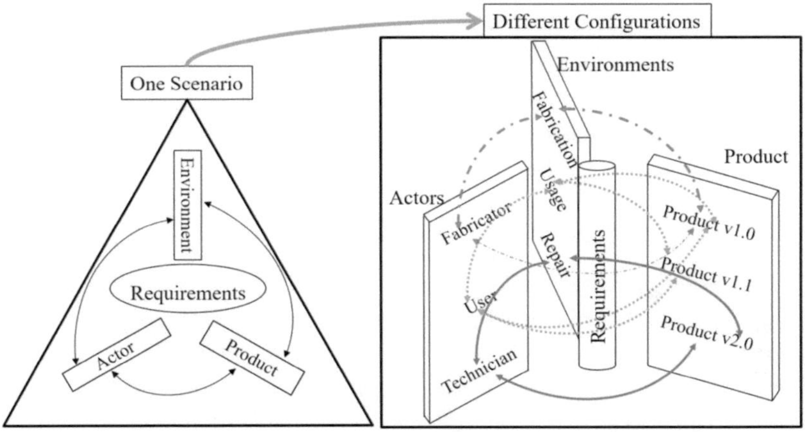

Figure 4-5: Efficient configuration of VR-model [Mah17a]

Similarly, if a database of different sub-models from different life-phases of a product is considered, the sub-models can be reused to form different VR configurations consisting of different combinations of these sub-models. This reusability can indeed contribute to the reduction of the overall effort for the preparation of different VR-model configurations. These configurations can be used to evaluate the product and to validate its requirements in its later life-phases. As a result, the future interactions and behaviour of a product in later life stages can be evaluated at the design stage.

4.1.3 Summary

This section has presented the theoretical base of this thesis and the idea of division of a complete VR-model into sub-models comprising of product, actor and environment. The sole purpose of performing such a division is to achieve smaller, reusable modules as sub-parts of a complete VR-model. The concept of contextual evaluation of a product and its effectiveness is also discussed. The CPM/PDD approach is extended to incorporate the actor and environment model explicitly inside external conditions and an outlook on the construction of different VR-model configurations is provided. Although the idea of the division of the complete VR-model is very appealing in terms of the reusability perspective of VR-model and its contents, its implementation is rather a challenging task. This is because the aforementioned reusability, substitution and recombination of sub-model in different use cases require an independent and isolated description of individual sub-models. An isolated description is only possible by avoiding any direct dependencies between the sub-models. Furthermore, these individual models have to work together / interact with each other to form one use case. The interaction between these sub-models should be described generically. Based on these challenges, the next section develops the approach for the description of these sub-models and their interactions.

4.2 Approach for the description of VR-models

Regarding the challenges associated with the description of the aforementioned sub-models and the requirements mentioned at the start of this chapter, the programming based solutions are not suitable for the implementation. The programming based preparation of a VR-model depends directly on the capability of the VR-software at hand and offers very limited reusability (detail in 3.3). Besides, any preparation method that focuses on VR-software preparation e.g. visual scripting, flow diagrams etc. encounters the same limitation. Therefore, this thesis focuses on the idea of performing the functional/behavioural description of VR-model outside of the VR-software. It is also important that this implementation idea should not be something completely alien to the industry so that it makes relatively easy integration of VR into the current development process possible. Considering these aspects, MBSE with SysML as the modelling language is chosen for the description of sub-models. This choice can be justified through the following arguments:

- MBSE and SysML are not fully new to the industry and have already found their place in industry (see section 3.1)
- SysML is a general-purpose and standardised modelling language [Iso17]
- SysML can describe the behaviour of our sub-models independent from each other using its behaviour diagrams and the model elements (detail to follow in sub-section 4.3.3)
- SysML also possesses the capability of modelling the interaction between sub-models using its standard interaction elements i.e. ports in a generic way (detail to follow in sub-section 4.3.2)
- SysML models have already found their way towards integration with existing PLM and CAD systems employing mecPro2 and Fas4M projects (see section 3.1)

As any VR-model may contain geometrical and behavioural descriptions of virtual objects, a concept based on the use of SysML can be seen in figure 4-6. Besides the geometry of a sub-model, the structural model in SysML describes the model components/parts, their interaction with each other and with the outside world. The interaction points in SysML are modelled using the *SysML Ports*.

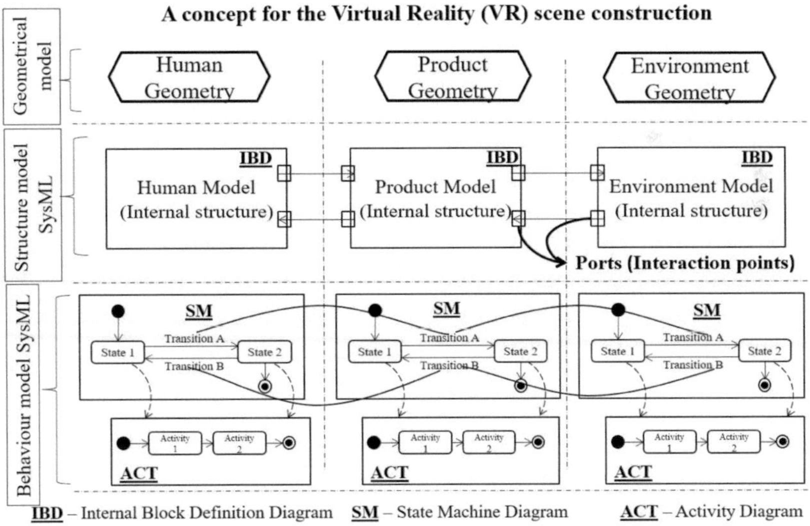

Figure 4-6: Concept of VR-model preparation [Mah17a]

The behaviour is modelled with the help of SysML behaviour diagrams and two of them are shown in figure 4-6 i.e. *State Machine Diagram (SM)* and *Activity Diagram (ACT)*. SM can model the states of a system with their individual behaviour. The transition between these states can occur as a result of the fulfilment of the mathematical condition(s) described in the model or due to a signal event from within the model. A signal can furthermore also come from outside the model. The behaviour of a state (e.g. *State 1* and *State 2* in figure 4-6) can be detailed by using activity diagrams that use control and object flows to implement behavioural aspects. The individual behaviour models can interact with each other by using structural description and the interactions defined over the sub-models' ports.

Based on the conceptual idea in this section, the next section presents the modelling approach for sub-models that contains the structural and behavioural descriptions of the individual models.

4.3 Modelling approach

To explain the modelling approach, a case example of a product is needed, so that the modelling process can be easily followed by the reader. A vacuum cleaner (manual) is not a very complex and at the same time, not a very simple product. Its use is intuitive, finds application almost in every household and requires a human user for its operation. Furthermore, the environmental model for its use phase e.g. a living room environment possesses moderate complexity. The cooperation of a vacuum cleaner manufacturer as an industrial partner of this research helps in acquiring the information related to the industrial development process. Therefore, a vacuum cleaner model is used as a case example to explain the modelling process and later used to build the first prototype in VR. The context of a vacuum cleaner in different life-phases can be seen in figure 4-7. This figure shows some of the possible actors and environments specific to the example of a vacuum cleaner.

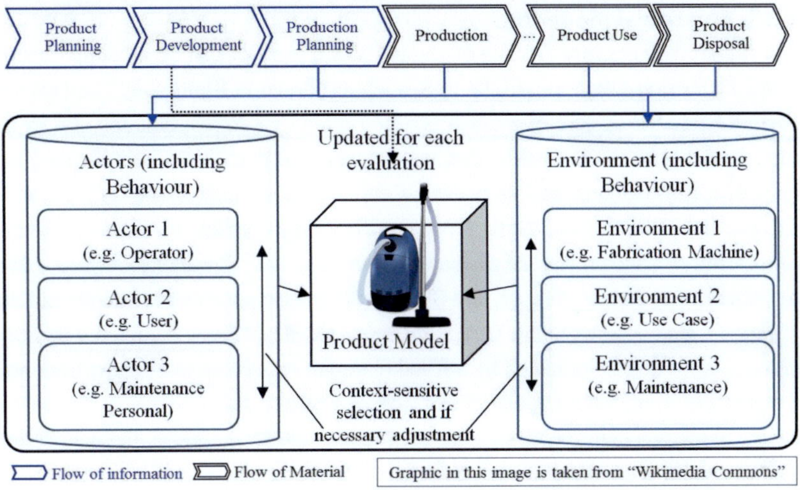

Figure 4-7: Context of a vacuum cleaner (Graphic[41]) [Mah17a]

It has already been mentioned in section 1.2 that the actor model development does not lie in the scope of this thesis; thus, the model of a *VR interaction device is considered as a proxy to actor model.*

The interaction device can be used by the VR-user to manipulate/use vacuum cleaner model in VR and its model description is developed in SysML. The vacuum cleaner model and a living room model considered as its usage environment are also modelled in SysML. The rest of this section will systematically present the modelling approach for these sub-models (i.e. interaction device, vacuum cleaner and living room). The first sub-section 4.3.1 describes the structural modelling of these three sub-models, followed by the definition of their interfaces in second sub-section 4.3.2. The third sub-section 4.3.3 describes the behavioural modelling process in details.

[41] Vacuum Cleaner Graphic : Wikimedia Commons: https://commons.wikimedia.org/wiki/Vacuum_cleaner#/media/File:Blue_vacuum_cleaner.svg [last accessed on 09.03.2020]

4.3.1 Structural modelling

The hierarchal structure and the owing parts of a product can be modelled inside *Block Definition Diagram (BDD)* in SysML as shown in figure 4-8. The vacuum cleaner being the system of interest is represented by a *Block*. A block is a structural model element in SysML that may represent a system, its components, the flow of information as well as materials etc. The components/parts of a *Vacuum Cleaner* are also modelled using blocks, as they may contain sub-parts or interaction definitions and are named as *Sensor_Module, Motor, Product_Interface, Handle, CableUnit, Power_Supply*, and *Wheels*. These components are linked with the *Vacuum Cleaner* block using *Directed Composition* relations. A directed composition relationship is indicated by a filled diamond connected to an arrowhead and conveys the structural compositions of different blocks. For example, in figure 4-8 *Motor* block is a component of *Vacuum Cleaner* block, therefore, the connection between them is of directed composition.

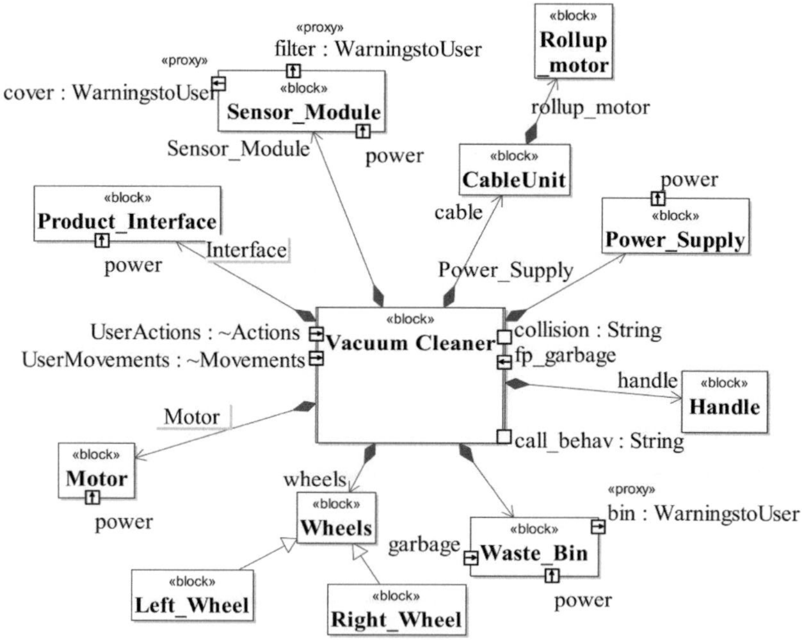

Figure 4-8: Vacuum cleaner structure as BDD

In this way, a BDD can represent a logical or physical decomposition of a system. Similar to *Vacuum Cleaner*, its component may also contain sub-components or parts that can be represented in separate BDDs. The *Wheels* block is connected with *Left_Wheel* and *Right_Wheel* block using a *generalisation relationship* (line connected with a hollow triangle). This indicates that the *Wheels* block is a more generalised form, whereas *Left_Wheel & Right_Wheel* are more specialised forms of *Wheels*. Small boxes on the edges of different blocks are the interface elements in SysML so-called "*Ports*" and will be discussed in next sub-section. The structure of a system can be described as detailed as needed and it may consist of multiple diagrams. However, the focus of this thesis is on the behaviour descriptions of sub-models and therefore, the structure model is developed just to as much detail as later needed for the description of behaviour models. Similar to the BDD of vacuum cleaner, a BDD for the environment model can also be constructed as shown in figure 4-9. The *Living Room* model may contain different objects like a sofa, table, lamp etc. that can be modelled accordingly.

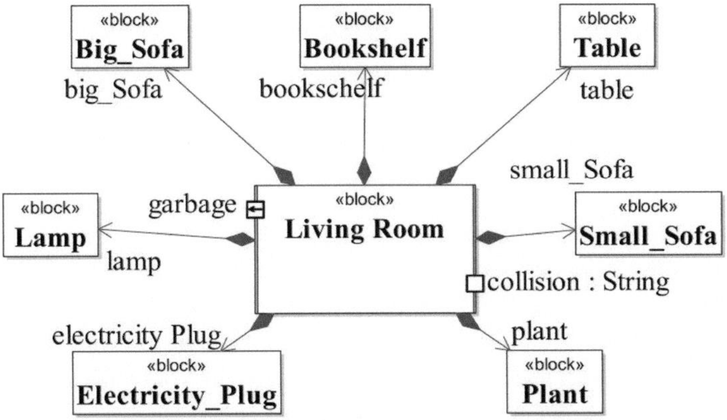

Figure 4-9: Living room structure as BDD

Similarly, the model of an interaction device can be built as a proxy for the actor model and can be seen in figure 4-10. A typical VR interaction device may have a few buttons and a joystick. The joystick itself is a multifunctional button that is a more generalised form of *Up, Down, Right & Left* movement buttons and hence, is linked with a generalisation relationship.

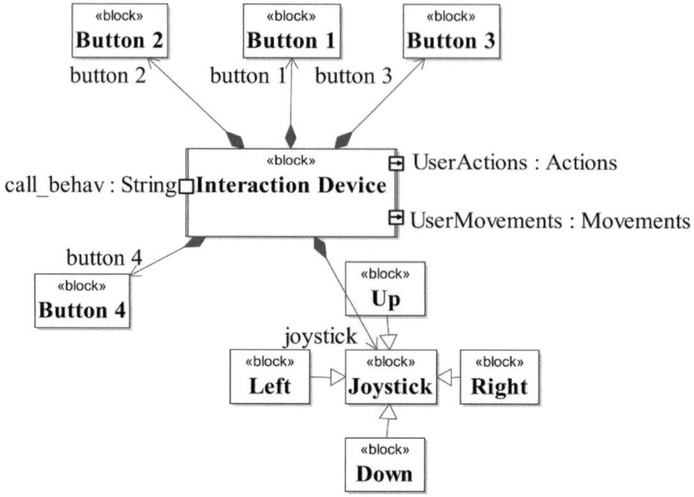

Figure 4-10: Interaction device as BDD

The sub-models that are described in this thesis mainly contain a primitive structural model that consists of only one BDD each. To have independent models, the structural definitions of all three sub-models are saved as separate project files. An important structural aspect of sub-models is the definition and connections of their interfaces. This is discussed in the next sub-section.

4.3.2 Modelling of the interaction

The last sub-section has provided an overview of the basic structural definition process and the resulting definitions are saved in separate project files. As a result, three project files are representing each sub-model individually. These sub-models have to work together and interact with each other to form one contextual use case of a product in VR, therefore, a new SysML project is needed that represents the overall system consisting of the three sub-models. This new project shall import and use the three projects containing so far only the BDDs of sub-models. Depending upon the desired functionality, these projects can be loaded as *read-only* or *read-write* accessibility options. In the case of *read-only* accessibility, the elements of the loaded project cannot be edited or modified directly in the current project, which is possible in the case of *read-write* accessibility. For instance, this can help to manage who can make changes in the shared project's content. The contents of the current model after loading the three projects (containing individual sub-models') can be seen as a package diagram (PKG) in SysML and also in the resulting model containment tree from the modelling tool (see figure 4-11). A *Package* in SysML is a model element that can contain any number of model elements or diagrams and is mainly used to achieve model organisation. It can also be understood as an analogy to a *folder* in windows based operating system. All three packages originating from the project files of the sub-models containing the individual BDDs and model elements. The sign of a "hand" in the containment tree indicates that the package is a shared resource.

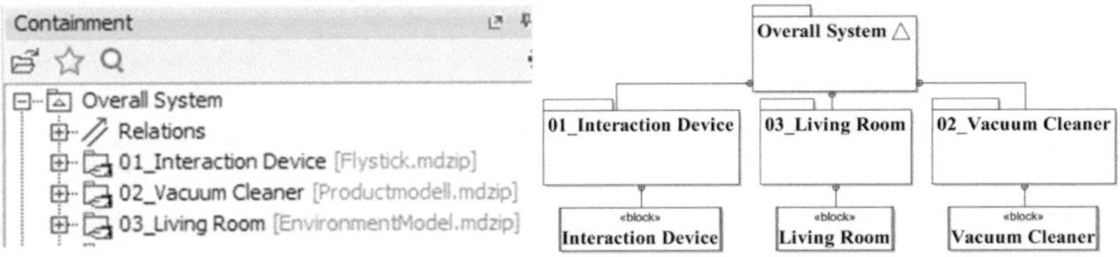

Figure 4-11: Containment Tree (Left) & Package Diagram (Right)

After loading the sub-models inside the new project, it is time to define information inside this new project (i.e. *Overall_System*) so that the interaction between the sub-models can be defined. As the overall VR-model consists of the three sub-models, it can be represented by modelling a representing block for the overall system (i.e. *Overall_System*) and connecting it with the blocks representing the individual sub-models using directed composition relationship. The resulting BDD can be seen in figure 4-12.

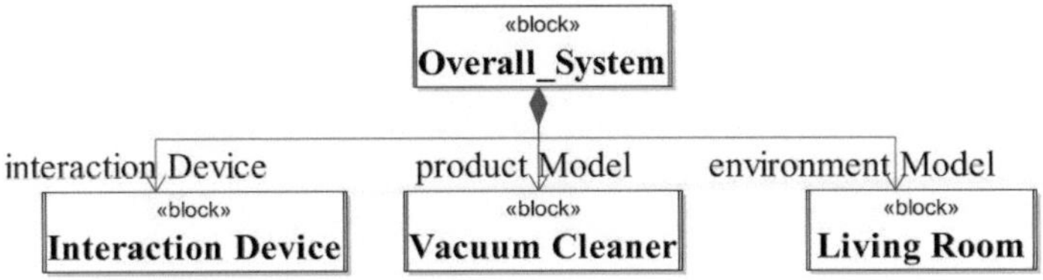

Figure 4-12: Structure of the overall system as BDD

After the definition of the overall structure, the next step is to define the interaction between these block as shown in figure 4-13. To better understand this figure, it is important to have a closer look at the ports already modelled in figure 4-8, figure 4-9 and figure 4-10. As the description of the three sub-models is loaded into the current project, the already modelled ports within individual sub-model descriptions are available for use. Two kinds of ports available in SysML i.e. full port and proxy port are defined as:

"SysML identifies two kinds of ports, one that exposes features of the owning block or its internal parts (proxy ports), and another that supports its own features (full ports)" [Omg18b]

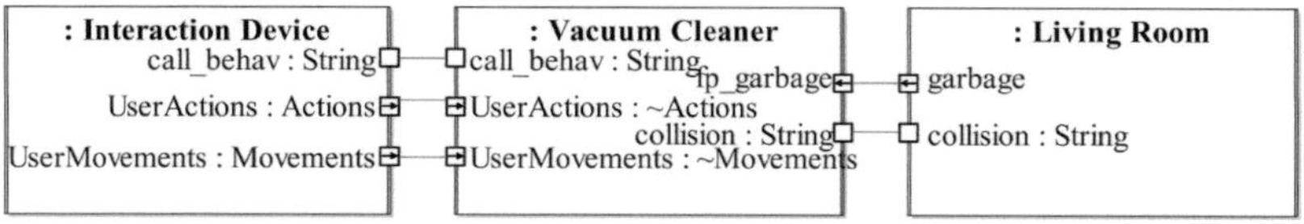

Figure 4-13: Internal structure as IBD (modified from [Mah18a])

Figure 4-13 is the internal block definition diagram (IBD) and shows the interconnection of the sub-models. The small rectangular boxes with and without the arrows inside them are the ports in figure 4-13 and are named as *UserActions, UserMovements, call_behav, collision, garbage* and *fp_garbage.*

SysML also offers a specialised block i.e. *Interface Block* that can be used to describe the functionality of a port. Interface block does not possess any behaviour related aspects or sub-parts, instead, it defines the information or material that flows through a port and specifies their direction e.g. signals, variables etc. For example, the *UserActions* port on *Interaction Device* is typed by the interface block *Actions* (figure 4-14).

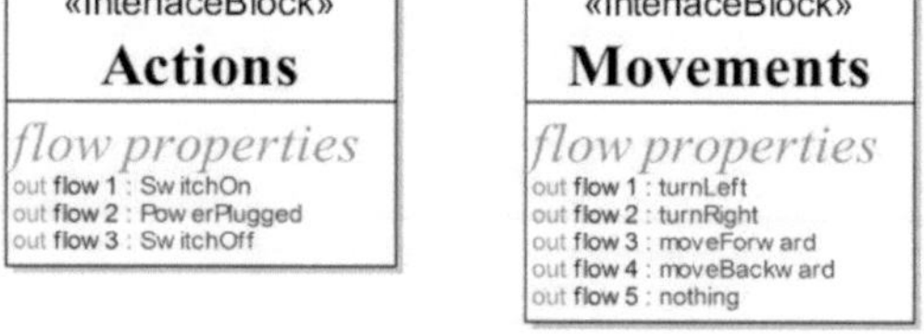

Figure 4-14: Interface blocks

The interface block *Actions* defines the actions that can be performed by the user on the *Vacuum Cleaner* i.e. plugging in the power and switching it on/off. *SwitchOn, PowerPlugged, SwitchOff* are signal elements of SysML and the word "*out*" inside the flow properties indicates that these signals flow out of *Interaction*

Device block. Similarly, the *UserMovements* port is typed by *Movements* interface block. The interface block *Movements* defines the possible movements performed by the user on the *Vacuum Cleaner* during its operation e.g. turning it, moving it forward or backwards etc. The same ports typed by the same interaction blocks are also present on the *Vacuum Cleaner* model, however, they indicate an inward information flow and the *conjugation sign* i.e. "~". For instance, the *UserActions* port on *Vacuum Cleaner* has the same properties as the one on *Interaction Device* block. The conjugation indicates that although the transferring information and the properties of both ports are the same, but the direction of information flow is the opposite. In other words, the information flows out of *Interaction Device* but flows inward to *Vacuum Cleaner*. The *garbage* port indicates that the garbage flows from the *Living Room* to the *Vacuum Cleaner*. During the usage of *Vacuum Cleaner*, there may be a collision taking place between the *Vacuum Cleaner* and an object inside the *Living Room* that is communicated to the *Vacuum Cleaner* model using the *collision* port. The *collision* port is typed by the data type "*String*" and it represents the name of the object taking part in the collision e.g. sofa or table etc. The *call_behav* port's functionality will be explained in the next sub-section.

In this way, the interaction between the sub-models can be modelled. The structural model can be further detailed to model the components within the sub-models. Furthermore, the interaction of these modelled components with each other can as well be modelled. However, for the sake of simplicity and to keep the focus on the development of the behaviour models in SysML, the structural model is only detailed to the level of utmost necessity. So far the last two sub-sections (4.3.1 & 4.3.2) have explained the structural modelling and the modelling of interaction, the next sub-section explains the modelling of the behaviour of each sub-model.

4.3.3 Behaviour modelling approach

The SysML behaviour diagrams can be used to describe the behaviour of the sub-models. These diagrams include different model elements to describe different aspects of system behaviour. The notations of most relevant model elements that are used in the explained modelling approach are described in table 15 in Annexure C along with a brief explanation of their use and functionality. As SysML is a very abstract language, there are multiple ways of describing one particular functionality. A model may use all the available behaviour diagrams or may only use one or two diagrams multiple times to achieve the designed functionality. In the description of the sub-models, state machine (SM) and activity diagrams (ACT) are used, as they allow the modelling of needed functionality and the remaining behaviour diagrams are not considered.

4.3.3.1 Product (vacuum cleaner)

The behaviour modelling process can be explained by having a look at the different states of a *Vacuum Cleaner* (product) as shown in figure 4-15. The black filled circle is the initial node and indicates the starting point of execution of an SM or ACT. Right after the start, the *Vacuum Cleaner* goes to a *PowerCheck* state where two activities i.e. *rec_sig_power* & *rec_sig_switchon* check if the power is plugged and *Vacuum Cleaner* has been switched on. Once both of these conditions are fulfilled the product goes into *inUse* state, where the *Usage* activity will be continuously performed unless or until a *Stopped* or *CoverOpen* signal comes in. *Stopped* signal refers to power being removed or *Vacuum Cleaner*'s operation being stopped and *Vacuum Cleaner* goes to the final node i.e. a solid circle contained inside a hollow circle. *CoverOpen* signal refers to the opening of the *Vacuum Cleaner*'s cover. The opening of the cover puts the *Vacuum Cleaner* into *ServiceNeeded* state and this state checks on the service procedures e.g. cleaning of filter or replacement of waste bin. As soon as the cover is closed, the *Vacuum Cleaner* goes back into the normal usage state.

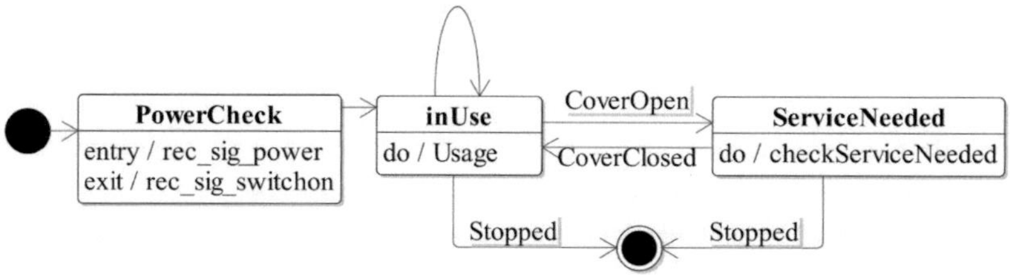

Figure 4-15: State machine diagram of product model (modified from [Mah17b])

Each state in figure 4-15 has its own detailed behaviour modelled in the form of activities. For instance, the *PowerCheck* state contains two activity diagrams i.e. *rec_sig_power* & *rec_sig_switchon* and are shown in figure 4-16.

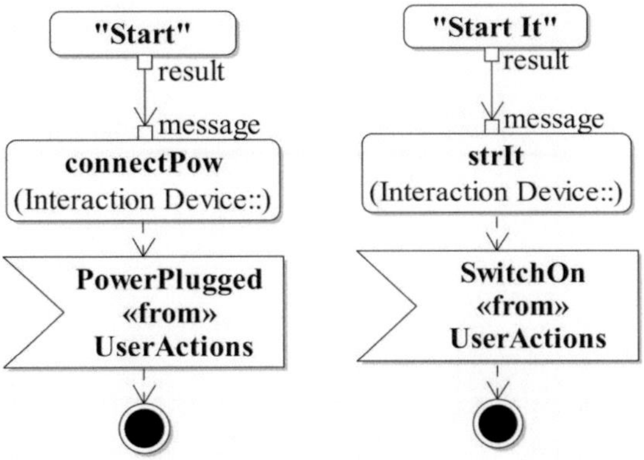

Figure 4-16: rec_sig_power (left) and rec_sig_switchon (right)

The rectangular boxes with round corners are the *actions* of an activity diagram. A wide variety of actions are supported in SysML that range from an internal behavioural action to calling another system to execute its behaviour or a sub-part of its behaviour. For example, *connectPow* and *strIt* are call behaviour actions that are calling for a certain aspect of the behaviour of *Interaction Device* to be executed. Both of these actions are followed by *accept event actions* i.e. *PowerPlugged* and *SwitchOn* respectively that are signals coming into the *Vacuum Cleaner* model over the port *UserActions* from *Interaction Device* model (see figure 4-13). This means that after initiating both actions the *Vacuum Cleaner* model waits for a feedback signal originating from *Interaction Device* model. The dotted lines indicate the flow of control and the solid lines indicate the object flow. Control flow can be typically understood as the flow of execution control and object flow as values being communicated from one action to another. Both of these activities complete their execution once *PowerPlugged* and *SwitchOn* signals are received. As a result, the *Vacuum Cleaner* enters into *inUse* state. The activity *Usage* being performed inside *inUse* (from figure 4-15) can be seen in figure 4-17.

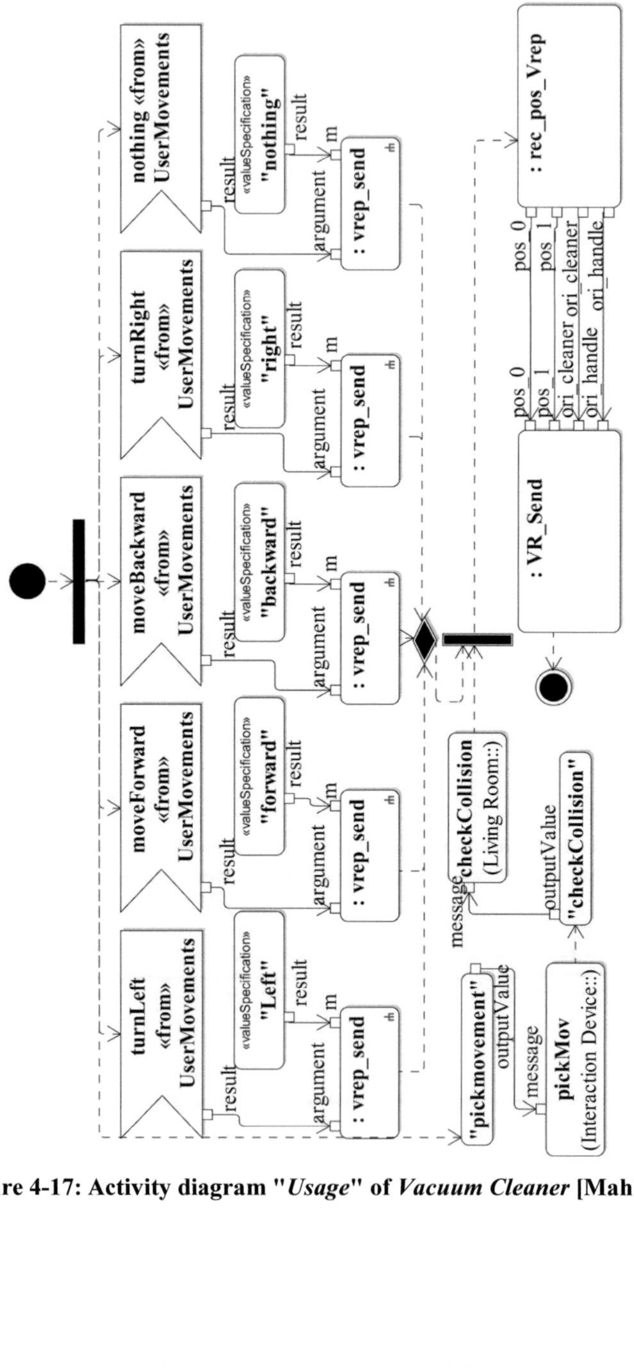

Figure 4-17: Activity diagram "*Usage*" of *Vacuum Cleaner* [Mah18a]

After the start of this activity, a *fork node (indicated as a black horizontal bar)* divides the control flow in multiple flows. This triggers all five of the accept event actions that are waiting for a signal on port *UserMovements.* Furthermore, it also activates the call behaviour action from *Interaction Device* i.e. *pickMov*.

As a result of the execution of *pickMov* (pickMov activity is explained in figure 4-21), one of the signals is received on the *UserMovements* port from the *Interaction Device* model. After the interpretation of the received signal (*turnLeft, moveForward, moveBackward, turnRight* or *nothing),* the relevant movement command is sent to a dedicated physics calculation software using respective *vrep_send* action. Once the movement command is transferred to the physics engine, the control flow arrives on the *join node (vertical black bar)* after passing through the *merge node (black diamond inside hollow diamond)* and waits for the second control flow coming from *checkCollision.* The sign "┌┼┐" on *vrep_send* action indicates that this action contains further diagrams for describing its complete behaviour. The integration of the physics engine will be discussed in subsection 5.2.1 and of VR-model in detail in section 5.3. As a result of any movements, there can be a collision taking place between *Vacuum Cleaner* and an object in the environment model. This is checked by calling the behaviour action *checkCollision* from the *Living Room* model. Once both the flows reach the merge node, the feedback from the physical engine about the new position, orientation, collision etc. values of *Vacuum Cleaner*'s objects are picked back by *rec_pos_vrep* action. The received feedback is evaluated and an update is sent to the VR-software using *VR_Send* action so that the contents of the VR-model and their properties can be updated. In this way, one execution of the *Usage* activity is completed.

It may occur during the execution of usage activity that the cover of the *Vacuum Cleaner* is open. In that case, the *Vacuum Cleaner* model moves to *ServiceNeeded* (figure 4-15) state and the *checkServiceNeeded* (figure 4-15) activity is performed. On the completion of this activity, the cover is closed and the *Vacuum Cleaner* model returns to *inUse* state.

4.3.3.2 Environment (living room)

Similar to the behaviour model of *Vacuum Cleaner* (product), the behaviour model of the *Living Room* can be developed and the state machine diagram for this model can be seen in figure 4-18.

Figure 4-18: State machine diagram of *Living Room*

The *Living Room* model keeps a check on the objects that can be present inside a living room and mainly checks if any of these objects are taking part in a collision. This model may also involve the positions and orientations of these objects and their individual behaviours. Each object has a collision value that can be 0 or 1. At the start of execution, this model sets the collision value for all the objects to default i.e. 0 indicating no collision and moves to standby from *setDefault* state using *setNone* activity.

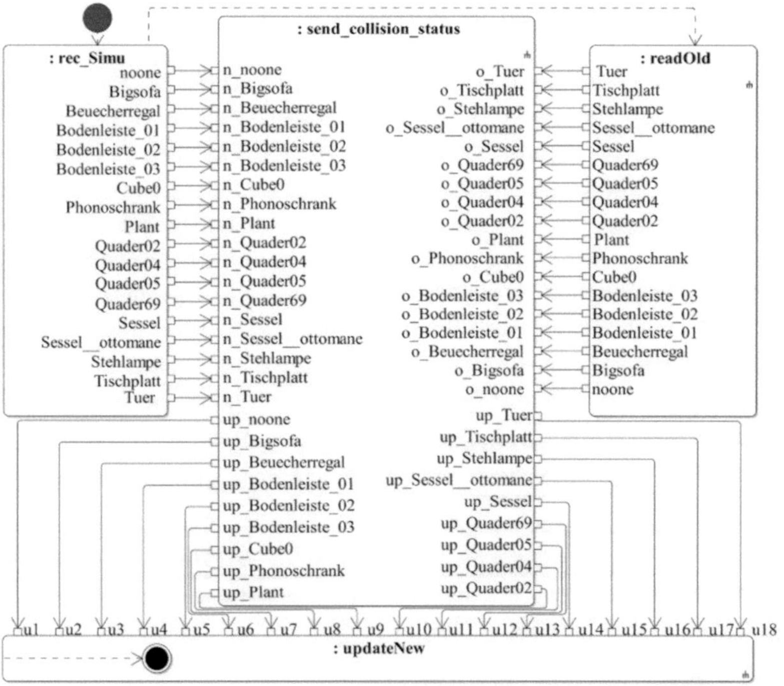

Figure 4-19: *checkCollision* activity from *Living Room* model [Mah18a]

Once the *Vacuum Cleaner* model calls the behaviour action *checkCollision* (figure 4-17) the activity of *checkCollision* (figure 4-19) behaviour of the *Living Room* is executed. After the start of the action, *rec_Simu* action picks the feedback from the physics engine regarding the objects' statuses. *Bigsofa, Buecherregal, Bodenleiste_01* etc. are the names of the objects in the living room environment. The received feedback is compared to *readOld* action that contains the status of the objects from the last execution run using *send_collision_status* action. As a result of this comparison, the information about the objects that have changed their collision state is extracted and communicated to the VR-software. Furthermore, the collision statuses of the objects are updated inside the model as well using *updateNew* action, so that they can be used for the next execution run. The actions *send_collision_status, readOld and updateNew* contain multiple activity diagrams to complete the functionality described above and therefore, contains the sign "┌┼┐".

4.3.3.3 Actor (interaction device)

Similar to the *Living Room* model, main & sub-behaviours of the interaction device are called by the *Vacuum Cleaner* model during its execution (*connectPow & strIt* from figure 4-16 and *pickMov* from figure 4-17). The interaction device also has only one state i.e. *inOperation* and stays on it. The *connectPow* (figure 4-20 middle) & *strIt* (figure 4-20 right) activities consist of a user action and as a result, a signal is generated that is communicated to the *Vacuum Cleaner* model using *UserActions* port. Both of these activities link to the *PowerCheck* (figure 4-15) of the *Vacuum Cleaner.*

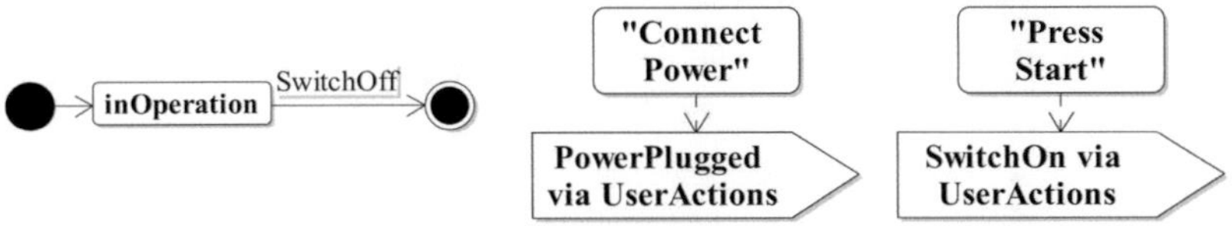

Figure 4-20: *Interaction Device* state machine (left), *connectPow* activity (middle) and *strIT* activity (right))

At any point during execution, the *Vacuum Cleaner* model can call the *pickMov* action, as a result, the *pickMov* activity (figure 4-21) from the *Interaction Device* model is executed. This activity mainly connects with the tracking system of the current VR-system over a User Diagram Protocol (UDP) connection.

The *rec_udp*[42] is an *opaque behaviour action* in SysML that allows direct integration of a script inside the modelling. For instance, *rec_udp* contains a small script that implements a UDP connection with the tracking systems and picks the current values of the physical interaction device. Depending upon the vendor of SysML tool, such a script can be written in multiple languages e.g. Java, Jython, JavaScript, Maple etc. The use of opaque behaviour action offers an excellent possibility to reduce modelling effort by packing a small chunk of object-oriented code inside a script. The opaque behaviour action *rec_udp* takes the IP address of the tracking computer as a string value and port as an integer value on which the tracking data is made available. After establishing the connection, *rec_udp* receives the current status of the interaction device from the tracking system. The received values are post-processed to extract the current values of the joystick, its coordinates, orientation and button values etc. An analysis of this data makes it possible to extract the user action that is eventually communicated in the form of a respective signal to the *Vacuum Cleaner* model so that user action can be realised.

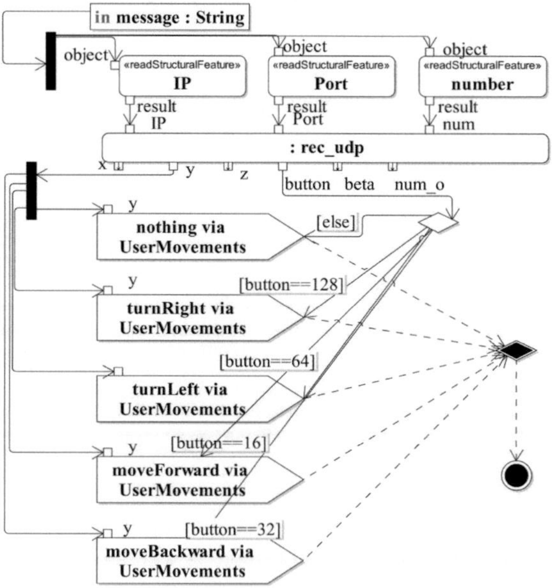

Figure 4-21: *pickMov* activity [Mah18a]

42 Code in Annexure A

Section 3.1 mentioned a few of the efforts to simulate the behaviour models of SysML. However, these methods rely on transformations in executable languages or Petri Nets etc. A detailed discussion or implementation of a transformation based method for behaviour model simulation are full-fledged topics for separate research. Therefore, instead of implementing a simulation methodology, the built-in functionality of the modelling tool is used to simulate the behaviour models. This simulation logic is based on the *Semantics of a Foundational Subset for Executable UML Models (fUML)* [Omg18a] and *State Chart XML (SCXML)* [W3c20] standards.

4.3.3.4 Summary

This sub-section provides the modelling approach for describing the behaviour of sub-models in SysML. The interaction between the sub-models and the execution of their behaviour are also explained in detail. Although this sub-section has provided the first proof of concept for modelling of sub-models' behaviour and their interaction over a port, the presented method still has limitations in terms of the speed of execution. Figure 4-17 shows that the *Vacuum Cleaner* model controls the overall execution process. After starting its own execution, it first initiates the behaviour of *Interaction Device* and waits for the feedback. After receiving and implementing the received feedback, it initiates the behaviour of *Living Room* model and waits for its execution to complete. Such an execution builds a sequential execution architecture containing unnecessary delays in the execution of the *Vacuum Cleaner* model itself. Furthermore, the initiation of the behaviours of the other two sub-models inside *Vacuum Cleaner* model leads to direct dependencies and the sub-models are not described fully isolated from each other. Therefore, the next sub-section attempts to overcome these limitations by developing a parallel execution architecture that eliminates unnecessary execution delays and also attempts to avoid direct dependencies in-between sub-models.

4.3.4 Sequential and parallel execution architectures

The behaviour modelling method described in the last section leads to a sequential execution architecture containing unnecessary delays. This sequential architecture can be visualised employing figure 4-22. As the *Vacuum Cleaner* model depends on the values/parameters of interaction device, the *Vacuum Cleaner* model calls the behaviour of interaction device while using the respective port right after the start of its execution and waits for feedback.

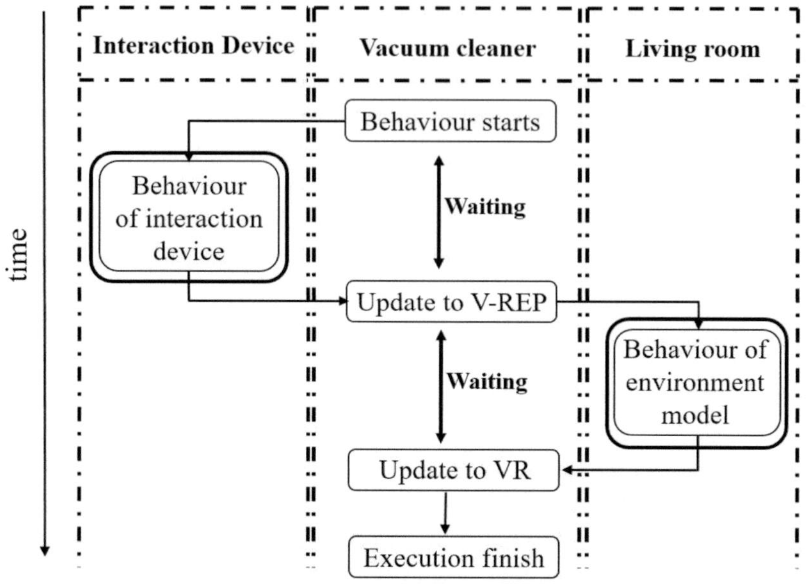

Figure 4-22: Sequential execution architecture [Mah18b]

After receiving the feedback from the *Interaction Device*, *Vacuum Cleaner* model updates the physics engine model (V-REP discussed in detail in sub-section 5.2.1) and calls the behaviour of the *Living Room* model. The *Living Room* model checks the properties of environmental objects inside the physics engine and in case of a collision event, it is communicated to VR-software. Again the *Vacuum Cleaner* model waits for the execution of *Living Room* model to complete and later, sends the updated *Vacuum Cleaner* model parameters to VR-software. In such an execution architecture, it is clear that the overall execution time of the *Vacuum Cleaner* model depends on the execution time of *Interaction Device* and *Living Room* model. As a result, unnecessary delays are incurred and slower execution of the *Vacuum Cleaner* model is carried out. A slower execution of *Vacuum Cleaner* models means that the updates sent to the physics engine and the VR-software will be taking place at a slower rate. If the physics calculations are not directly performed inside the VR-software, a smooth simulation in VR directly depends on the rate at which the updated parameters are sent to VR-software. A slower execution of the *Vacuum Cleaner* model refers to a slower rate of parameter updates to VR-software that eventually refers to a slower product simulation in VR.

The VR simulation obtained using this execution architecture showed non-smooth movements of the *Vacuum Cleaner* model in VR and noticeable steps in the movements could be observed. Although figure 4-22 depicts the problem at hand, an experiment is carried out to quantitatively examine the scope of the problem and to understand the delay in the execution in a better way. This experiment records the overall execution time of all three models and the times between two adjacent data transfer to VR-software. A commercial SysML behaviour model execution tool is used along with the physics engine (V-REP). The start times and the end times of the executions of each sub-model are recorded in the form of a text file. The execution times for each sub-model can be seen in figure 4-23.

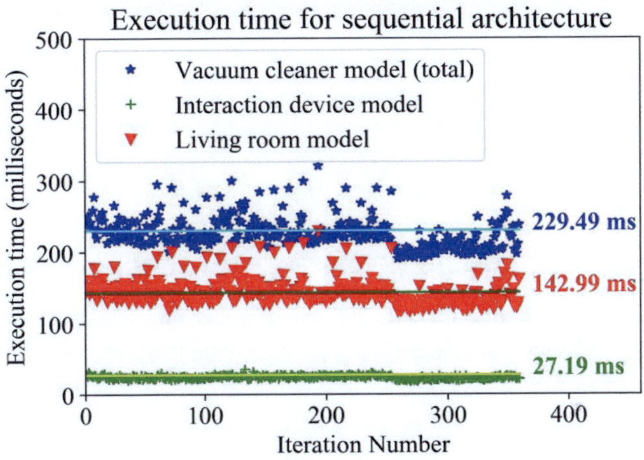

Figure 4-23: Model execution times for sequential architecture [Mah18b]

It can be seen that the execution time (on average 229.49 milliseconds (ms)) of the *Vacuum Cleaner* model is highest among all sub-models. This is because the *Vacuum Cleaner* model contains the execution times of *Interaction Device* and *Living Room* models in addition to its own execution time, as it is waiting idly during their executions.

As the *Vacuum Cleaner* model and the *Living Room* model send parameter updates to VR-software, it is important to note down the times between two adjacent data transfers made from *Vacuum Cleaner* and *Living Room* models each to VR-software. Therefore, the time between two adjacent data receptions on the VR-software side from *Vacuum Cleaner* and *Living Room* model are recorded individually.

Furthermore, it is supposed that the *Vacuum Cleaner* and *Living Room* model only send one parameter update to VR during their one execution cycle so that a clean comparison can be achieved. The recorded data transfer times are shown in figure 4-24.

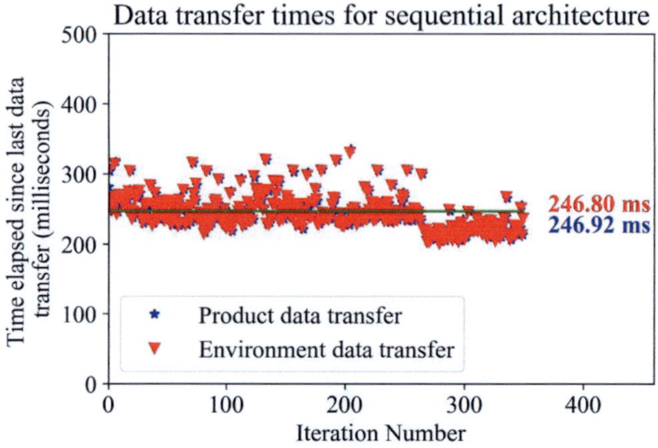

Figure 4-24: Data transfer times for sequential architecture [Mah18b]

The average transfer time for both *Vacuum Cleaner* and *Living Room* model is approximately 247 ms. As the execution of the *Living Room* model (142.99 ms) is much faster than that of the *Vacuum Cleaner* model (229.49 ms), it can be expected that its data transfer time will be lower than that of the *Vacuum Cleaner* model. However, the *Living Room* model has roughly the same data transfer time as the *Vacuum Cleaner* model. This high data transfer time is due to the sequential execution architecture and the fact that each model has to wait for its turn during the execution (see figure 4-22). Thus, the sequential execution architecture causes a decrease in the rate at which the parameters' updates are sent to VR-software even for the sub-models with faster execution. Furthermore, the second disadvantage of a sequential execution is the direct dependency of the product model on the other sub-models for its execution which eventually constitute a negative aspect regarding the reusability of these sub-models. Therefore, the modelling approach described so far needs improvement and a new execution architecture is needed. The new execution architecture should support the paralleled execution of sub-models and avoid unnecessary delays in the execution of any of the sub-models.

Parallel execution of sub-models can be obtained; if no sub-models initiates the execution of any other sub-models and all three sub-models are executed independently. This can be achieved by initialising the behaviours of each sub-model by the overall model (*Overall_System* in figure 4-12) that contains these three sub-models. This requires the redefinition of the interfaces as well as the behavioural model of *Vacuum Cleaner*. Therefore, the ports for the communication between the sub-model are modified and figure 4-25 represents the modified version of figure 4-13. The newly added elements are colour filled for easy identification.

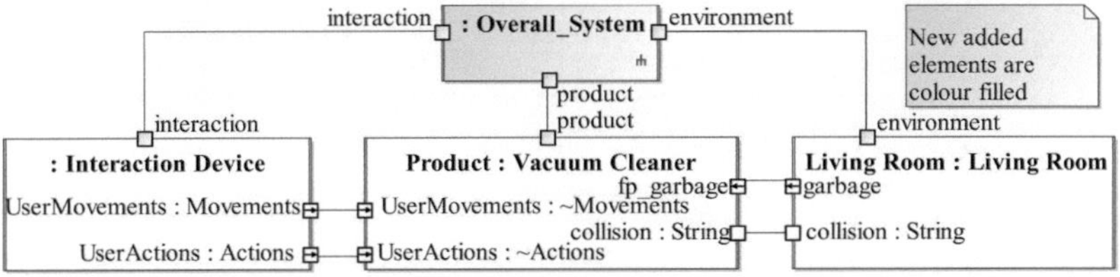

Figure 4-25: IBD for parallel execution [Mah18b]

The main modifications involve the removal of the *call_behav* port that was used by the *Vacuum Cleaner* model to call the behaviour of *Interaction Device* model and the incorporation of six new ports. Out of the six newly added ports, three are added on the *Overall_System* and one on each of the sub-models. These new ports allow direct interactions between *Overall_System* and the sub-models. These direct interactions can be utilised to call the behaviour of sub-models independent from each other and can be realised by an activity diagram as shown in figure 4-26. This figure consists of three independent and endless execution loops. The first loop corresponds to the behaviour of *Interaction Device*, second to the *Vacuum Cleaner* and the third to the *Living Room*.

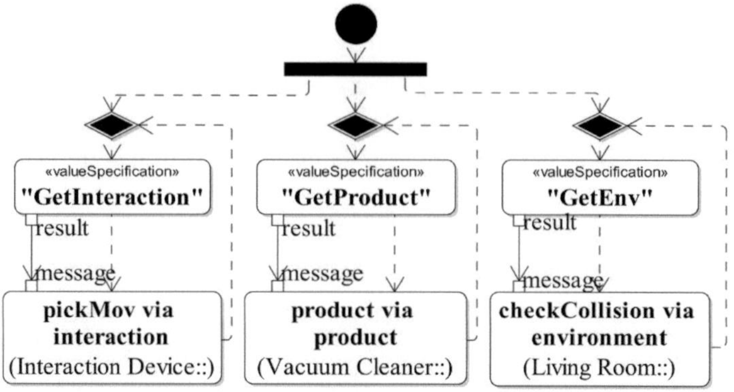

Figure 4-26: Parallel execution of sub-models [Mah18b]

After redefining the ports on the sub-models, the execution can be started by execution of *Overall_System* that contains the activity diagram (figure 4-26) as its main behaviour. As a result, the *Overall_System* uses the three new ports (*interaction*, *product* and *environment*) to initiate the behaviour of sub-models individually. This eliminates the need for *Vacuum Cleaner* to call the behaviour of other sub-models and thus, the activity "*Usage*" (from figure 4-17) of the *Vacuum Cleaner* model also needs modification. The call behaviour actions *pickMov* and *checkCollision* are no longer necessary, as the product model will no longer initiate the behaviour of other sub-models. Hence, *pickMov* and *checkCollision* are removed and the resulting activity diagram can be seen in figure 4-27.

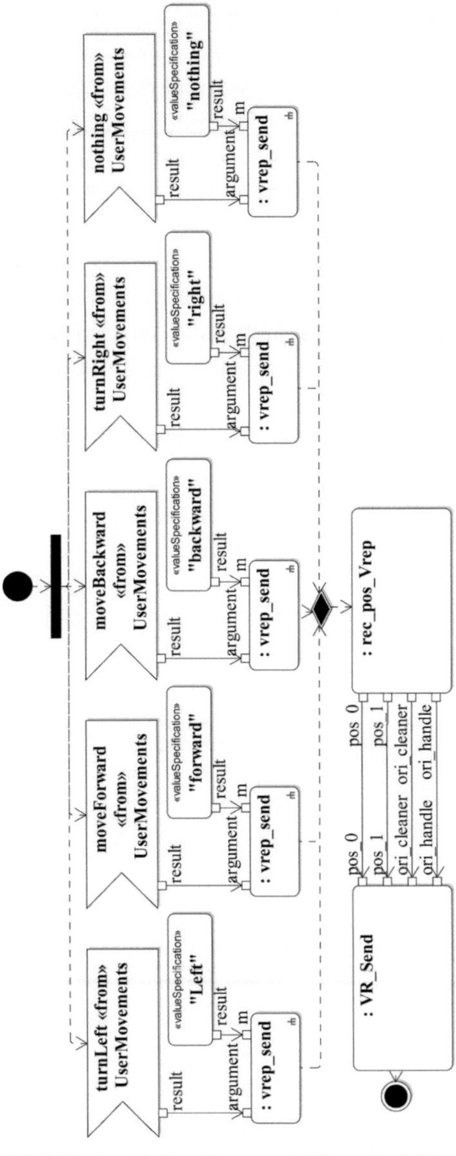

Figure 4-27: Modified activity diagram "*Usage*" of *Vacuum Cleaner*

Figure 4-26 shows that the individual behaviour models are executed in parallel and none of the sub-model can cause a delay in the execution of others. This can be again quantitatively verified by recording the execution times of all sub-models. Figure 4-28 shows a clear improvement in the execution time of the *Vacuum Cleaner* model i.e. 84.34 ms against the old value of 229.49 ms. However, it can be noticed that the execution times of *Interaction Device* and *Living Room* model have increased considerably. The reason for this increase is the sharing of the overall processing power among three parallel execution loops (execution threads) corresponding to each sub-model. This can be handled eventually by the use of parallel computers and by executing each sub-model on a separate computer. However, the detailed discussion of parallel computing is not included in this thesis.

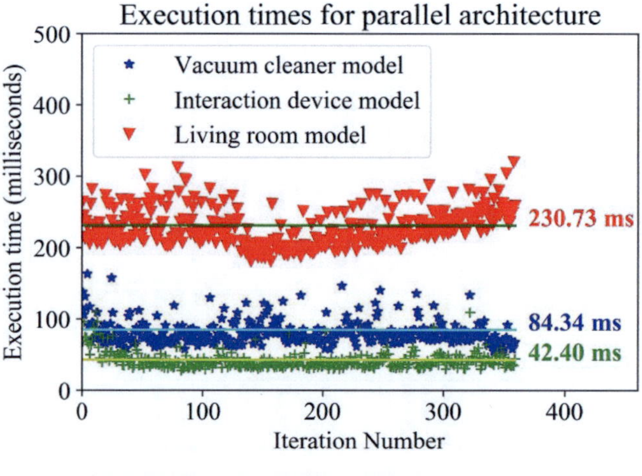

Figure 4-28: Execution times for parallel architecture [Mah18b]

The main goal of the development of a new parallel execution architecture is to achieve a faster transfer of parameter updates to VR-software from SysML behaviour descriptions. The new data transfer times to VR-software are recorded and are shown in figure 4-29. It can be seen that the data transfer rate from the *Vacuum Cleaner* model to VR-software has almost increased by a factor of two as compared to the sequential execution architecture (that was on average approximately 247 ms). This means that the updates about the parameters of the *Vacuum Cleaner* model are sent two times faster to VR-software that eventually leads to a smoother and faster simulation of the *Vacuum Cleaner* model in VR. There can be no major reduction observed in the data transfer time for the *Living Room* model and it corresponds to the higher execution time of the *Living Room* model i.e. on avg. 230.73 ms. The *Living Room* model mainly updates the collision statuses and the positional parameters of objects in the environment. These updates may well be less critical as compared to a *Vacuum Cleaner* model that is continuously in motion. Furthermore, in an active VR simulation, the parameters of the *Vacuum Cleaner* model are continuously communicated to VR and the parameters of the living room are communicated in case of a change in the collision status or object positions/orientation. A lower transfer rate for the *Living Room* model can thus be accepted as a trade-off.

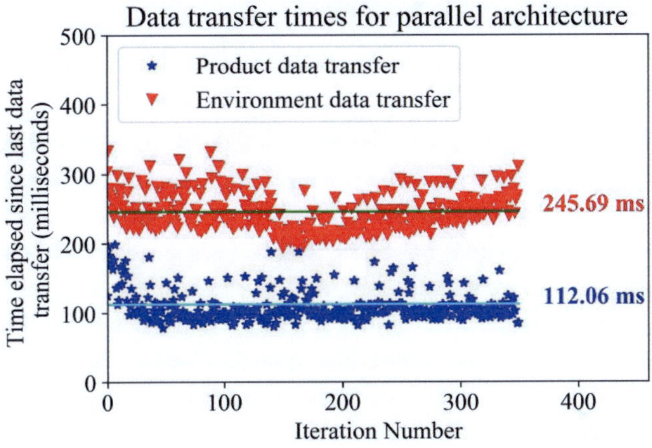

Figure 4-29: Data transfer times for parallel architecture [Mah18b]

The new parallel execution architecture not only reduces the unnecessary delay in the execution of the *Vacuum Cleaner* model but also removes direct dependencies within the sub-models. Each sub-model can now execute independently from other sub-models that is very helpful as far as the reusability perspective of these sub-models is concerned. However, the substitution/replacement of any sub-model will require additional modelling effort, as the interfaces (in figure 4-25) and the activity diagram (figure 4-26) for parallel execution will require remodelling. Therefore, the presented approach is further refined to eliminate the need for remodelling by using SysML instances. This is elaborated in detail in the next sub-section.

4.3.5 Automatic behaviour model initialisation and use case generation

The behaviour modelling methodology presented so far enables the parallel execution of sub-models to form a product use case. Also, the parallel execution is achieved by manual initialisation of each sub-model's behaviour. To achieve reuse of different sub-models and to form new use cases, it is important to replace/substitute any sub-model and to achieve automatic parallel initialisation of the behaviour of sub-models. Although the replacement/substitution is achievable with the presented methodology, it requires remodelling effort. For instance, the replacement of the product model from the parallel execution architecture (figure 4-25 & figure 4-26) requires

- the current execution to be stopped,
- remodelling of figure 4-25 and figure 4-26 for a new use case and
- start of a new execution.

This might seem a relatively easy and intuitive task for the explained example of *Vacuum Cleaner*, however, it can be prone to error in case of a large and complex overall SysML model. Therefore, in this section, the use of *SysML Instances and the concept of inheritanc*e is presented. SysML instances can help to create dynamic configurations of VR use cases, as they eliminate the need for additional modelling for each use case and allow automatic initialisation of the behaviour of sub-models in parallel. This can be achieved by developing a *High Level Solution Architecture (HLSA)* that is the core model for each solution architecture against the given problem [Ale18]. For instance, an HLSA model can contain the generic structure of a system and its components, its interfaces, flows etc. that can be reused in the development of its sub-systems. In this way, HLSA serves as a reference model to avoid inconsistencies in the modelling of sub-systems. This new approach does not modify the behaviour model of any of the previously described sub-models, but focuses towards a more generic definition of the interfaces between the sub-model and thus modifies the structural models only.

Figure 4-12 can be modified to represent a more general structure of the complete model as HLSA. Figure 4-30 shows the HLSA for the system consisting of three sub-models i.e. product, actor and environment along with the representation of their ports. This architecture should be generic enough to cover all the possible use cases that may exist. Therefore, all three sub-models should possess all the ports and interface definitions that any of its versions may possess. Only then a replacement of any of the sub-model's version with another version can be made possible.

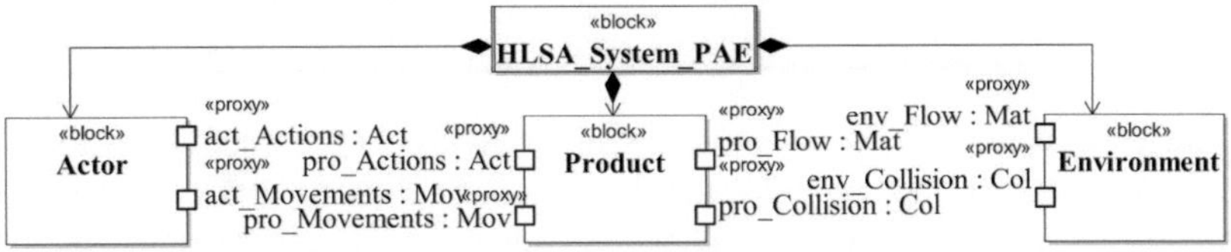

Figure 4-30: System Architecture (HLSA) [Mah19b]

The use of the HLSA model and the definition of all possible interface points inside it is useful in the construction of different variants of the system. This may lead to models with redundant information (150 % model[43]), however, this can help manage a complex system and its variants [Aki19].

The *Overall_System* block from figure 4-12 is replaced by *HLSA_System_PAE* and its internal structure can be seen in figure 4-31 which shows the connection and interfaces between different sub-models that are used by these sub-models to communicate data or information. As the goal is to avoid manual initialisation of the behaviour of each sub-model, this figure does not include any connections or information exchange between sub-models and the *HLSA_System_PAE* model as was previously done in parallel execution architecture. The interaction between the sub-models mainly consists of the flow of signals, variables, materials etc. and all these elements should also be defined in this structure. At this point, all the elements that may flow from one sub-model to others in any possible use case should be modelled. Now the generic structure of the overall system, the interfaces and the flows between the sub-models are defined. As a next step, this complete model can be saved as a separate project file for later use in the development of the sub-models. This project so far contains only two diagrams i.e. the BDD from figure 4-30 and IBD from figure 4-31.

[43] The concept of 150% model refers to the idea of [Aki19]. It means that the model is readily equipped with excessive information about system and its interface, that may not be needed in all application scenarios. However, this redundant information may well be very helpful in building different variants of a systems as discussed by [Aki19].

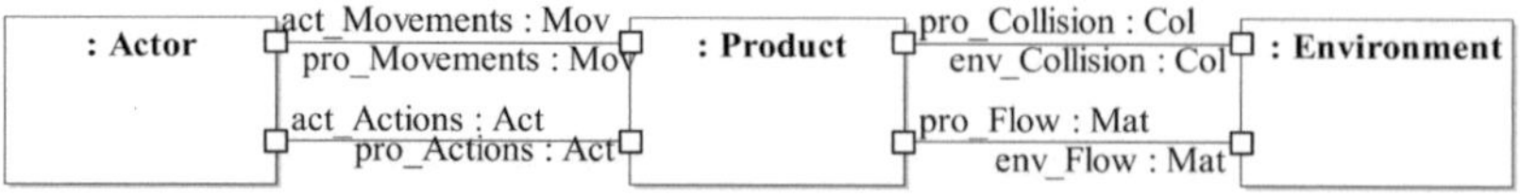

Figure 4-31: IBD of HLSA_System_PAE [Mah19b]

The three sub-models of product, actor and environment modelled so far inside HLSA do not contain any behavioural descriptions and only define the generic structure of the overall system saved inside a project (*Project file 1* in figure 4-33). Now this project file can be reused to detail the structure and behaviour of the sub-models. The structure as well as the behaviour descriptions of sub-models shall use the HLSA and the interface definition defined inside it. As a result, consistent model interfaces can be achieved in all the sub-models and all the possible use cases. These consistent interfaces can be explained employing the BDDs in figure 4-32. Considering that the actor model may contain a model of interaction device or a virtual actor (e.g. human model of a domestic user) depicted as blocks named *InteractionDeviceV1* and *DomesticUserV1* in figure 4-32. These two models represent two variations or versions of the *Actor* model. Both these blocks are linked with *Actor* block from *HLSA* model using generalisation relationship in SysML. The generalisation relation is denoted by a solid line connected to a hollow triangle and represents inheritances of constrains. This means that *InterationDeviceV1* and *DomesticUserV1* block inherit the constraints defined inside the *Actor* block. These constraints are mainly the interface, ports and signals defined inside the *Actor* block. For example, the ports on *InteractionDeviceV1* are the direct replica of the ports of *Actor* block as they are inherited ports indicated by the sign "^" on the port. It is important to save both these BDDs in separate project files (*Project file 2& 3* in figure 4-33) so that they may facilitate the reuse of sub-models. After creating individual projects for each version of the *Actor* model, the behaviour of the *InteractionDeviceV1* and *DomesticUserV1* can be modelled while using the same ports and interfaces inherited from the *HLSA* model. This modelling of the behaviour will take place on the same lines as already described in sub-section 4.3.3.

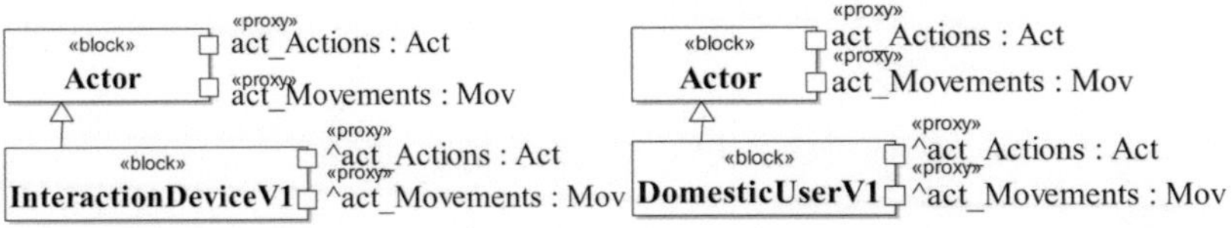

Figure 4-32: BDD for *InteractionDeviceV1* and *DomesticUserV1* [Mah19b]

Similar to the *Actor* model, the models of *Product* and *Environment* may also contain multiple versions. For instance, if the product and environment also possess two versions each, the overall structure of the complete system can be depicted as figure 4-33. *VacuumCleanerV1* and *VacuumCleanerV2* are the two versions of the *Product* model whereas *LivingRoomV1* and *FabricationEnvironmentV1* are the two versions of the *Environment* model. Each of these versions has to be saved in individual project files and as a result, there are a total of seven project files (*Project file 1-7* in figure 4-33). One project file containing the description of *HLSA* (*Project file 1* in figure 4-33) and is reused in the six projects (*Project file 2-7* in figure 4-33) that are containing different versions of sub-models. In this way, the interfaces and the ports on all the sub-models will be consistent and compliant with that of *HLSA*. These versions of sub-models can be described in detail to include behavioural descriptions in the same way as described in sub-section 4.3.3.

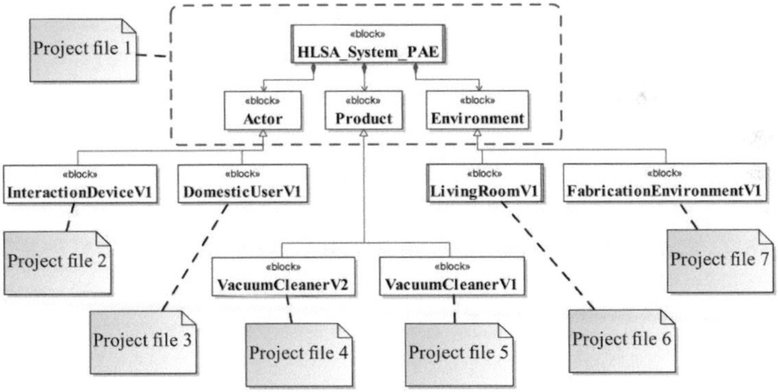

Figure 4-33: Overall structure and components of HLSA [Mah19b]

Once all six versions include their behavioural descriptions, the next step is to use these different versions to construct different use cases. This can be done by first loading all six versions of sub-models inside *HLSA* project file and as a result, the model containment tree from modelling tool will contain all seven models (see the left side of figure 4-34).

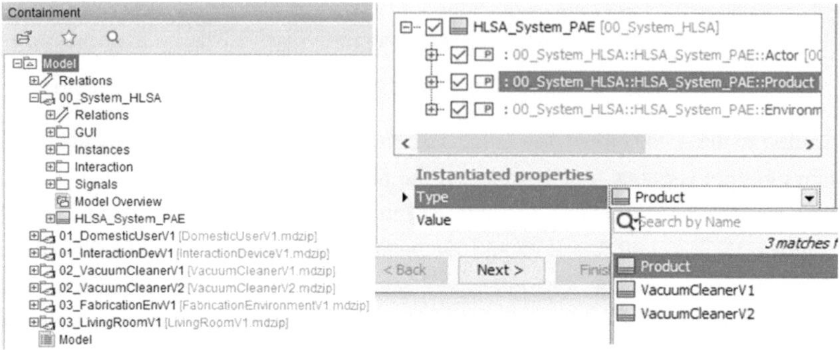

Figure 4-34: Model containment tree (left) and automatic instantiation dialog box (right) [Mah19b]

At this point, the instance specification functionality of SysML can be used to create one instance of the *HLSA_System_PAE* model. Figure 4-34 (right) shows the instance specification dialogue box from the modelling tool that offers the possibility to choose one version for each of the sub-models. As a result, an instance specification can be created e.g. an instance of *VacuumCleanerV1* inside *LivingRoomV1* being controlled with the help of *InteractionDeviceV1* is shown in figure 4-35. This instance shows that *HLSA_System_PAE* consists of three parts (i.e. sub-models) and also specifies the exact version of each part used in the construction of this particular instance. Execution of this instance automatically initialises the behaviours of all sub-model in parallel which are involved in the current instance. As a result of this execution, a real-time VR simulation (detailed explanation in chapter 5) representing one use case of the product in VR is obtained (see figure 4-35 (right)).

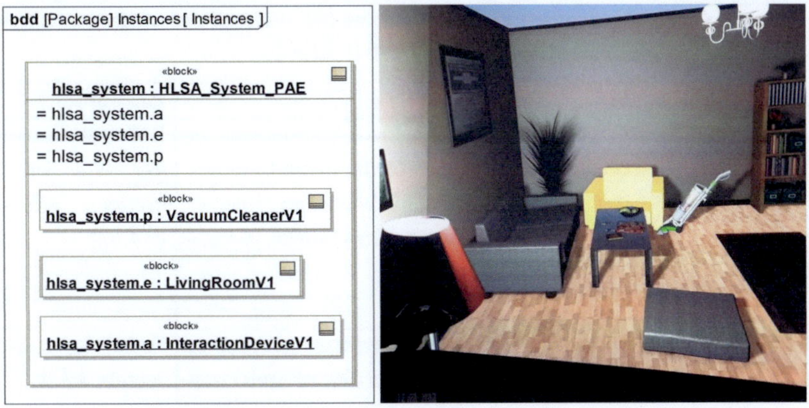

Figure 4-35: One Instance (left), VR simulation (right) [Mah19b]

Similarly, different instances can be created easily with a few mouse clicks to create different use cases containing a different combination of the sub-models. In this way, the remodelling needed for the creation of use cases while using parallel execution architecture (see sub-section 4.3.4) can be avoided. Furthermore, the use of instances and generalisation help to keep the interfaces on all the sub-models consistent that is eventually very helpful for the reuse/substitution of sub-models. Based on the complete modelling approach presented in this chapter until now, generic guidelines can be formulated and are presented in the next sections.

4.4 Guidelines

This chapter has described in detail the modelling methodology for describing the behaviour of sub-models that resulted from the methodological division of the complete VR-model. Section 4.3 has described the modelling approach in detail and discussed the advantages and shortcomings of different architectures. A dynamic VR-model resulted from the use of a SysML behaviour model which encouraged the reuse of parts of the overall VR-model. The knowledge about modelling presented in this chapter can be summarised in the form of general-purpose guidelines for the behavioural description of a dynamic VR-model. These guidelines can be described as follows:

Table 4: General-purpose guidelines for the behavioural description of a dynamic VR-model in SysML

	#	Purpose	Explanation	Examples
1. Identification of the system of interest	1.1	System decomposition	Identify the main components or sub-systems of the overall system, its interface, sub-systems, interaction between sub-systems, signals and flows	Refer to figure 4-7 and subsections 4.3.1 & 4.3.2
	1.2	Interfaces	Identify all interfaces relevant to each possible version of each component/sub-system. Must include interfaces for internal and external communication.	Refer to subsection 4.3.2
	1.3	Information exchange	Identify all possible signals or parameters or values that can flow between sub-systems or with the outside world	Refer to figure 4-14 and subsection 4.3.2
2. Structural models	2.1	General architecture (structure)	Create an HLSA model containing the generic structural definitions of the overall system and its sub-systems	Refer to subsection 4.3.5, figure 4-30 and figure 4-31
	2.2	General architecture (interaction)	Model the already identified signals, interaction points and flows identified in 1.2 & 1.3 inside HLSA from 2.1	Refer to figure 4-30, figure 4-31 & figure 4-14

3. Dependencies and model management	3.1	Sub-systems' versions	Realise all possible versions of each sub-system of the overall system	See sub-sections 4.3.1 to 4.3.3
	3.2	Managing project files	Each version of each sub-system should be saved and organised as a separate project file	Refer to figure 4-33
	3.3	HSLA as the reference model	All project files (3.2) shall link with their respective generic representation in the HLSA model with generalisation relationship	Refer to figure 4-32
	3.4	Consistent interfaces	Inherit the ports, signals and flow specification for each version of each sub-system from their generic representation from HLSA model	Refer to figure 4-32
4. Behaviour models	4.1	Behaviour modelling	The behaviour model for each version of the sub-system shall be modelled inside the respective project file (3.2)	Refer to sub-section 4.3.3
	4.2	Behaviour modelling (dependencies)	The behaviour model description of a sub-system shall be kept independent from that of other sub-systems	Refer to sub-section 4.3.3
	4.3	Interaction over ports	The information exchange or interaction between two sub-systems shall only take place over the interfaces (ports in SysML) defined in the HLSA model	Refer to figure 4-25 and figure 4-33
	4.4	Changes in interfaces, signals and flows	Is there a redefinition or modification in any of interfaces, signals or flows needed? It shall be performed inside the HLSA model	Refer to figure 4-30 and figure 4-31

	4.5	Missing interfaces, signals or flows	A new interface, signal or flow element is needed? it shall be modelled inside the HLSA model	Refer to figure 4-30 and figure 4-31
5. Execution /use case generation	5.1	Execution/ simulation	The behavioural model of the overall system or its sub-systems shall not be initialised manually. If done otherwise, additional modelling effort is needed.	Refer to figure 4-17, figure 4-26 and figure 4-35
	5.2	Execution/ use case generation	The automatic initialisation of behavioural models shall be achieved by using instance specifications & inheritances in SysML. This enables parallel execution of sub-systems and avoids additional modelling effort	Refer to figure 4-35

The developed guidelines can be categorised in a total of five categories as shown in table 4. Table 4 contains the purpose and brief explanation of the mentioned guidelines as well as reference to example sections and figure to this thesis for the detailed explanation. After identification of the system of interest, the modelling process starts with the modelling of structure. After the modelling of structure, the important aspects regarding the independent and isolated description of sub-model along with the indication of model management aspects are mentioned. The fourth part lays out the important aspects related to the modelling of behaviour models and finally, the last part talks about automatic parallel execution of sub-model and generation of different use cases of the product. These guidelines are intended for use by the individuals in academia as well as in industry who are interested in using SysML models for VR simulations.

4.5 Summary

This chapter has presented the methodological backbone of this thesis in detail. Starting from the identification of requirements associated with a new method for the preparation of VR-models at the start of this chapter, a basic contextual evaluation model of a product in VR is presented in sub-section 4.1.1. The idea of the division of a complete VR-model into sub-models of product, actor and environment is put forth in sub-section 4.1.2. Furthermore, the advantages of performing such a division and the challenges associated with its implementation are discussed. An approach based on MBSE and the use of SysML is adopted to describe the sub-models. The primary factors that resulted in SysML being the choice as the description language are discussed in section 4.2. The process of modelling the structure as well as the behaviour of the sub-models in SysML is explained in detail and in a systematic way employing a case example of a vacuum cleaner (section 4.3). A living room environment along with a model of interaction device is considered as the context of the vacuum cleaner. The first modelling idea is based on a sequential execution architecture. Although this architecture can realise a VR simulation, it incurs unnecessary delays and direct dependencies in-between sub-models. These problems are addressed by presenting a parallel execution architecture that removes the unnecessary delays and executes all sub-models in parallel. An additional modelling effort is needed in the construction of each new use case while using parallel execution architecture. To avoid this extra modelling effort, the presented methodology is further extended by incorporating the use of SysML instances and the concept of inheritance.

Thus, this chapter has systematically developed and presented the modelling method for the description of VR-models. The final result of the presented method enables the reuse of the behaviour of sub-models described in SysML. The use of instances in SysML eliminates the need for remodelling effort and the sub-models' behaviour can be automatically executed in parallel. The research question 1 (see section 3.5) has been answered by performing the division of complete VR-models in sub-models of actor, product and environment. The possibility to construct different use case scenarios of a product in VR by recombining these sub-models achieves reusability of the parts of the complete VR-model. The research question 2 (see section 3.5) has also been answered by creating isolated behaviour descriptions of the sub-models that do not possess any direct dependencies with each other.

Once the isolated behaviour descriptions of sub-models have been developed, the next step is to integrate these behaviour models with VR and the physics simulation tool so that they can drive an active VR simulation. This integration method is presented in the next chapter.

5 Simulation of product properties in VR

After presenting the modelling methodology in the previous chapter, this chapter explains the overall process to achieve a simulation in VR while using the behaviour models developed in SysML. The overall process involves the development of the visual model for VR, physics simulation model and its integration within SysML. Figure 5-1 shows the final goal of this chapter in the form of information exchange between different tools involved in achieving the VR simulation. *UserInterface* can contain a graphical user interface or simply be an interaction device for VR. The input from the user interface/interaction device flows into the *Models and Descriptions* in SysML (explained in chapter 4). SysML models communicate with and control *V-REP (Physics Engine)* simulation and the updated properties of the virtual objects are communicated to *VR-software*. *VR-software* finally projects the updated content onto a *CAVE* type VR-system and the VR-user can experience an interactive product simulation in VR.

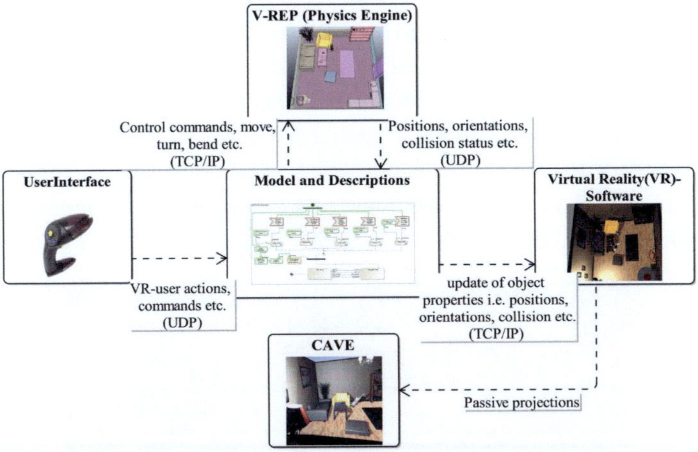

Figure 5-1: Exchange of information between different tools during simulation [Mah18a]

The VR-software contains the visual model in the form of virtual objects. The aspects related to the development of virtual objects corresponding to the product and the environment models are discussed in sub-section 5.1. To keep the overall implementation easily understandable, the case example of the vacuum cleaner inside the living room environment is taken to explain the overall process. The visual model for the case example is developed accordingly. Next, the incorporation of physical calculations is brought to light. The aspects related to the choice of physics engine and physics model development are discussed in sub-section 5.2. Furthermore, the use of SysML behavioural descriptions to control the physics engine is discussed in detail. In sub-section 5.3, the integration of SysML behaviour descriptions with VR will be discussed that result in an example simulation of product use case in VR. Based on the knowledge gained during the implementation, a generic information flow during simulation in different VR-systems is presented in sub-section 5.4. Finally, sub-section 5.5 provides a summary of this chapter and an outlook on the use of the presented method in different VR-systems.

5.1 Visual model development for VR

The VR-software contains the visual model of the product and the environment. As this thesis uses an interaction device as a proxy for the actor model, the visual model of the actor is not included here.

5.1.1 Living room visual model

The living room model used in this thesis originated from a media project conducted in the Media Production Group at Technische Universität Ilmenau [Ang08]. The goal of this media project was to study and evaluate the applicability of virtual prototypes for use in the product development process. A scenario based on a vacuum cleaner model inside a living room was evaluated. The living room model developed in this study can be seen in figure 5-2. [Ang08]

Figure 5-2: Screenshot of the VR-scene [Ang08]

This living room environment consists of the common objects that can be found in any domestic living room environment e.g. sofa, table, bookshelf, TV etc. The size of the virtual living room is 25 square meters and the roof is placed at a height of 2.70 meters. At the time of the development of this living room environment, *Virtual Design 2 (VD2)* was used as the VR-software. VD2 was originally developed by the company *vrcom GmbH* that was later merged in IC.IDO and VD2 was not developed further as a product. In the development of the living room environment, a mixture of different software tools was used with a focus on the development of a photorealistic model. Mainly, CAD software, 3ds Max[44] and VRML [Vrm20] exchange format were used to develop the final model in VR. Furthermore, to generate different textures, Photoshop[45] was used as well. [Ang08]

For the examination inside this thesis, the same living room model[46] is used inside RTT Deltagen as VR-software with slight modifications. These modifications mainly involved

- the light settings,
- objects' materials and textures,
- rearrangement of object hierarchical tree and
- removal of unnecessary virtual objects from the VR-scene.

5.1.2 Vacuum cleaner visual model

The product model is usually developed as a CAD model during the design and therefore, the CAD model is used to prepare the visual model of the product in VR. The CAD model for the vacuum cleaner[47] is taken from an online website[48] that is available in the form of SolidWorks parts and assembly files. In the CAVE type VR-system available at Technische Universität Ilmenau, RTT Deltagen is used currently as the VR-software. Therefore, based on the information presented in

---

[44] 3DS Max: https://www.autodesk.de/products/3ds-max/overview [last accessed on 09.03.2020]

[45] Adobe Photoshop: https://www.adobe.com/de/products/photoshop.html [last accessed on 09.03.2020]

[46] The conversion from IC.IDO to RTT Deltagen was performed by Dr. Helge Drumm (technical supervisor of the FASP in Technische Universität Ilmenau)

[47] Vacuum cleaner CAD model: https://grabcad.com/library/vacuum--3 [last accessed on 09.03.2020]

[48] Grabcad.com: Gradcad content is available for personal, non-commercial and internal use as mentioned under "3. License" under https://grabcad.com/terms [last accessed on 09.03.2020]

table 1 and figure 3-10, VRML is the best option available for transferring the geometrical model from CAD to VR because

- it carries the most information (surfaces, colours, size, textures etc.) about the original model compared to other CAD exchange formats,
- SolidWorks supports the direct export in VRML format and
- RTT Deltagen supports the direct import of VRML format.

The VRML format can transfer the information about the geometries, scale, colouring etc. However, one limitation of the VRML format is that it does not carry any information about the object/part hierarchy from the original CAD model. Therefore, after importing the VRML file inside RTT Deltagen, the imported model does not contain any information about the object/part hierarchy and each object/part is imported as a single surface. The hierarchy tree for the vacuum cleaner model is built by manually grouping/rearranging the objects inside VR-software. It is important to build the hierarchy tree inside VR-software so that the positions, orientations, rotations of objects of interest can be read and manipulated in VR-software as desired later during the simulation. For instance, rotating a parent object (owing object in the hierarchy) in VR-software will cause all the subparts/child objects to rotate accordingly. After performing the manual modifications, the resulting VR-scene containing a vacuum cleaner inside a living room model can be seen in figure 5-3.

Figure 5-3: Screenshot from the VR-scene in RTT Deltagen

So far the VR contents only consist of visual objects and objects' transforms (positions, orientations etc.) without any information about animation or simulation. The example hierarchy tree inside RTT Deltagen containing the virtual objects can be seen in figure 5-4 (left). The transform of any particular object and object's properties can also be visualised and manipulated inside RTT Deltagen. For example, figure 5-4 (right) shows the transform and properties of an object cleaner that represents the virtual model of the vacuum cleaner. The actual position and the orientation of any virtual object are present inside its transform and can be manipulated as well. Furthermore, the objects can be assigned visible or invisible in the object's properties in VR-software.

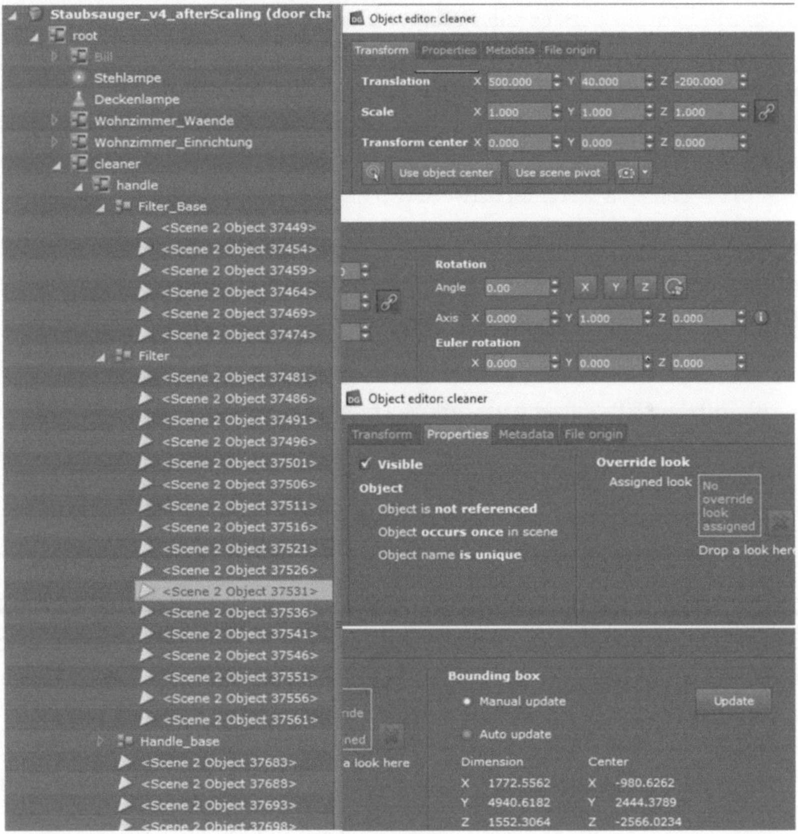

Figure 5-4: Hierarchy tree (left), object transform and properties (right)

The basic interaction possibility inside CAVE type VR-system is facilitated with the help of an interaction device e.g. a flystick. A flystick allows the user to change the viewpoint, move or fly inside the VR-scene etc. and in this way, the virtual objects can be visualised from different points of view. Furthermore, it is also possible to hide certain objects/parts to achieve visualisation of internal objects/parts.

This visual model is used to exactly create the physics simulation model that is explained in the next sub-section.

5.2 Calculation of physical behaviour

In order to verify the product's functionality or dynamic behaviour, it is very important to perform physics calculations on the product during its virtual model evaluation. Therefore, to build the behaviour of a product model in VR, it is of utmost importance to perform real-time physical calculations on the geometric objects as well. For instance, the very basic behaviour of a product may need the physical calculations to examine the reaction of certain forces on a certain component or the collision detection between two objects.

A few of the currently available VR-software products offer dedicated physics calculation modules, however, the use of these integrated modules is subjected to the following limitations (see also sub-section 2.2.3 and section 3.3):

- Their use is subjected to extensive programming or visual scripting etc.,
- different VR-software use different programming languages that lead to interoperability issue between them and
- as different VR-systems use different VR-software, the use of integrated modules for performing physical calculations limits the reusability of VR content across different VR-systems.

To overcome the above-mentioned limitations, this thesis proposes the use of an external dedication physics calculation software. In this case, the VR-model has to be constantly updated with the newly calculated parameters from the physics engine to update the virtual objects' transforms/properties in real-time. As the description of the dynamic behaviour of the VR-model is done outside VR-software (i.e. in SysML), the use of physics calculation software external to VR-software helps to keep the overall description of VR-model independent of the used VR-software. A VR-software independent description of a VR-model also possesses the potential of being reusable across different VR-systems (also see the reasoning in section 3.3). Furthermore, performing physical calculations on a product with the complex geometrical model can be very demanding in terms of the required computational power and it is also critical that it should function in real-time. Carrying out the physical calculations outside of VR is helpful as it can let us perform the computations on a computer other than the one used for rendering the contents of a VR-scene.

There are several physics engines currently available as open-source and as commercial licences. A few of the renowned physics engines are

- Bullet Physics[49],
- Open Dynamics Engine[50] (ODE),
- NVIDIA PhysX[51],
- Newton Dynamics[52] and
- Vortex Studio[53]

A detailed comparison of a few of these physics engines and some additional ones is provided by [Hum12] who examine them for different parameters like speed, precision, parameter tweaking etc. These physics engines are usually available either in form of a software package or build libraries. Furthermore, there are a few software solutions available that allow the construction of a geometric scene and integrate multiple physics calculation engines for the user's choice. For example, two of such software solutions exist as robotic simulators i.e. V-REP [Vre20] and Gazebo [Koe04]. Gazebo is in general open-source whereas V-REP offers a free version for academic use only. The significant features of both these simulators can be summarised as described in table 5. The most powerful feature of V-REP is its API support for multiple programming languages and hence, it can attract a

[49] Bullet Physics: https://pybullet.org/wordpress/ [last accessed on 09.03.2020]
[50] Open Dynamics Engine: https://www.ode.org/ [last accessed on 09.03.2020]
[51] NVIDIA PhysX : https://developer.nvidia.com/gameworks-physx-overview [last accessed on 09.03.2020]
[52] Newton Dynamics: http://newtondynamics.com/forum/newton.php [last accessed on 09.03.2020]
[53] Vortex: https://www.cm-labs.com/vortex-studio/ [last accessed on 09.03.2020]

broad audience. The external APIs are used to control the objects inside the simulators and typically, their positions, orientations, collisions status etc. can be manipulated and read as well.

Table 5: Gazebo vs V-REP

	V-REP	Gazebo
Open-source	No (but free for academic research)	Yes
Supported physics engines	Bullet 2.78 & 2.83 ODE Vortex Newton	Bullet ODE DART[54] Simbody[55]
Main Language	LUA	C++
External APIs	C/C++, Python, Java, Urbi, Matlab/Octave	C++

A comparative analysis of both of these simulators for use in the Robot Operating System (ROS)[56] is presented by [Nou14] that concludes:

- V-REP is more intuitive, user-friendly and contains more features
- Gazebo is more integrated into ROS framework, is open-source but requires extra tools to achieve functionality comparable to V-REP
- Gazebo is more hardware-demanding than V-REP

On the basics of intuitiveness, support for multiple external APIs and relatively less expensive execution of VREP, V-REP is chosen as the simulator for performing the physical calculation on virtual objects in VR. The next sub-section will explain in details the development process of V-REP model.

5.2.1 V-REP model development

V-REP typically contains a virtual scene of simplified geometric shapes. The scene in V-REP must be the replica of the VR-scene, so that the updated parameters after the physics calculations may well be communicated to update the objects in the VR-scene. For example, the positions, orientations and sizes of objects in V-REP

---

[54] Dynamic Animation and Robotics Toolkit(DART): https://dartsim.github.io/ [last accessed on 09.03.2020]

[55] Simbody – Multibody Physics API : https://simtk.org/projects/simbody/ [last accessed on 09.03.2020]

[56] ROS – Robot Operating System : https://www.ros.org/ [last accessed on 09.03.2020]

must correspond/ be proportionate to that of the VR-scene. By default, V-REP can support the geometric import of the following file formats:

- Mesh formats
  - .STL[57] – Contains the geometry in the form of triangle meshes
  - .OBJ[58] – Wavefront OBJ contains polygonal data in ASCII form
  - .DXF[59] – Drawing exchange format
- Model formats
  - COLLADA[60] – XML based schema for the transportation of 3D assets
  - SDF[61] – XML based format for defining objects and environments in robot simulators
  - URDF[62] – XML based format for representing a robot model

To develop the environmental model in V-REP, the living room objects from the VR-scene (see figure 5-3) are exported as individual STL files and later are imported in V-REP. Although this import generates a living room environment proportionate to that of VR containing simplified geometric meshes in V-REP, the orientation is not the same. The difference in the orientation (see figure 5-5) is the result of different global coordinate systems in RTT Deltagen and V-REP. Therefore, the orientation of the environment model inside V-REP has to be adjusted manually.

---

57 STL Format : http://www.fabbers.com/tech/STL_Format [last accessed on 09.03.2020]
58 OBJ : http://www.fileformat.info/format/wavefrontobj/egff.htm [last accessed on 09.03.2020]
59 DXF Format : http://www.fileformat.info/format/dxf/egff.htm [last accessed on 09.03.2020]
60 COLLADA - https://www.khronos.org/collada/ [last accessed on 09.03.2020]
61 SDF : http://sdformat.org/spec?ver=1.6&elem=geometry [last accessed on 09.03.2020]
62 URDF - Unified Robot Descriptions Format : http://wiki.ros.org/urdf [last accessed on 09.03.2020]

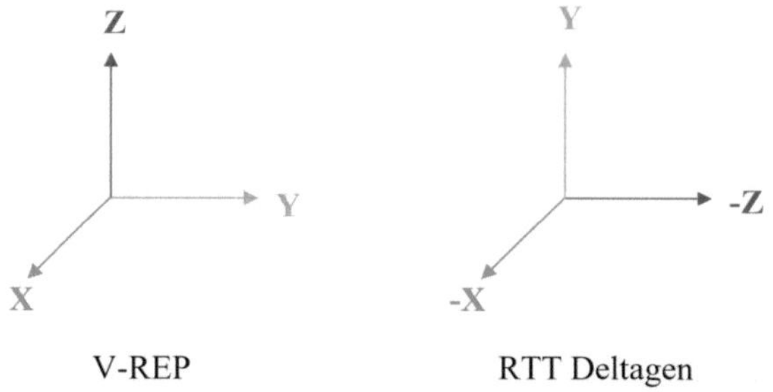

Figure 5-5: Coordinates system in RTT Deltagen and V-REP

As the STL file does not contain any kinematic information about joints, rotations, degrees of freedoms (DoF) of joint etc., it is not suitable for the transfer of vacuum cleaner model. The vacuum cleaner model contains wheels, handle joint, geometric shapes in between joints and is available as a CAD model in SolidWorks. The URDF format can contain the information about the joints, their position, the geometric meshes connecting the joints, information about coordinates etc. and therefore, offers the best possibility to transfer CAD model to V-REP. To transfer the vacuum cleaner model from SolidWorks to V-REP, the URDF-Plugin[63] that is available as an add-on for SolidWorks is used. This plugin provides an interactive GUI inside SolidWorks and lets the user build the hierarchy tree of the exported model manually. A part or assembly or multiple assemblies inside SolidWorks can be selected as the model base that represents the first object in the hierarchy (*base_link* in figure 5-6). This defines the first parent in the hierarchy and also specifies the number of children it contains. Similarly, each joint of interest can be identified and the geometric assemblies or parts between two joint nodes can be identified as linked geometric components.

[63] SolidWorks to URDF Exporter by Stephan Brawer: http://wiki.ros.org/sw_urdf_exporter [last accessed on 09.03.2020]

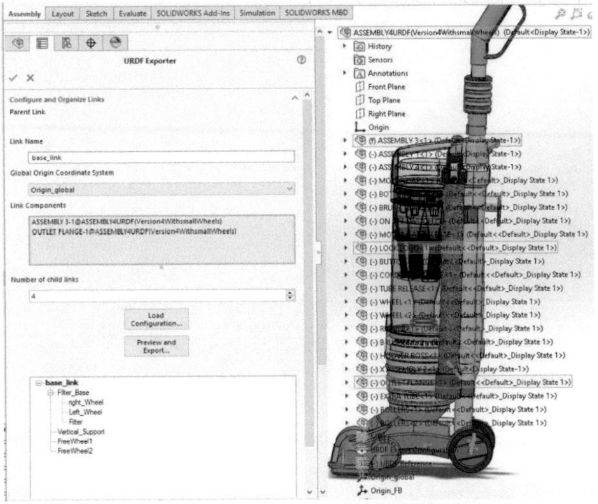

Figure 5-6: SolidWorks to URDF export

The URDF export from this hierarchy tree automatically detects the joints and their type (e.g. fixed, revolute etc.) as well as their exact positions and axis of rotations from the CAD model. The exporter calculates the exact placement of axis of rotation of a joint based on the assembly constraints defined inside SolidWorks. Before performing the final export the user can individually rename joints, view their axis of rotations, coordinate systems, orientation values etc. that is presented in the form of a dialog box (figure 5-7). The purpose of this dialog box is also to provide the user with the possibility to verify the information that is automatically extracted by the URDF exporter before performing the final export. The exported package contains a URDF file that contains the information about the complete model in the form of XML text and the geometries between two adjacent joints exported as STL mesh files. The XML file specifies all the nodes present in the model and specifies the name as well as locations of STL meshes that links two corresponding joints. Furthermore, the information about the joint type, its location, the axis of rotation etc. is also specified. A more detailed overview and guidelines to perform the export of SolidWorks assembly into URDF are also available in the form of an online tutorial[64].

---

[64] SolidWorks Assembly to URDF tutorial: http://wiki.ros.org/sw_urdf_exporter/Tutorials/Export%20an%20Assembly [last accessed on 09.03.2020]

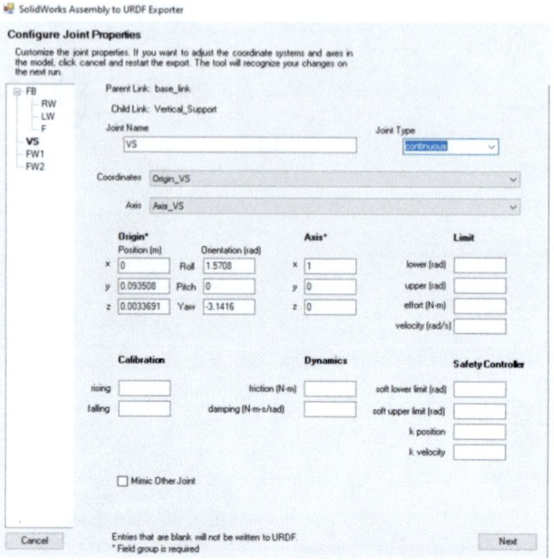

Figure 5-7: URDF exporter dialog box

The exported URDF model can be imported inside V-REP by using the default URDF importer and the resulting model in V-REP can be seen in figure 5-8. As a result, the exact geometrical model of the vacuum cleaner is constructed in V-REP with the help of URDF exporter. The imported model has preserved the information about the inertia, mass, joint locations, joint types etc. as well the hierarchy defined at the time of the export from SolidWorks.

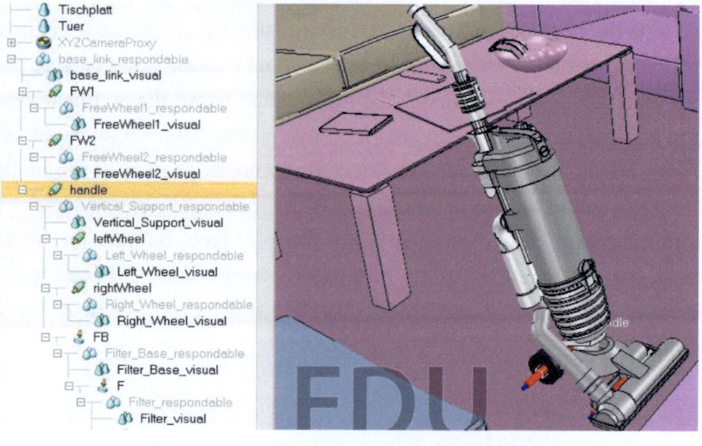

Figure 5-8: Vacuum cleaner in V-REP

This sub-section has described the process of creating the V-REP model that is the exact replica of the VR-scene (see Figure 5-3). Although the scene is a geometric replica, the geometric representations contain fewer amounts of polygons so that fast and real-time physics computations can be facilitated. This complete V-REP model with simplified geometrical objects can be seen in figure 5-9. The next step is to integrate this model with the SysML behaviour models presented in chapter 4. This is explained in detail in the next sub-section.

Figure 5-9: V-REP model for the computation of physics calculation

5.2.2 Integration of physics engine (V-REP) with SysML

In order to control the V-REP physics model from the behavioural descriptions in SysML, it is important to achieve bidirectional communication between the SysML models and V-REP. V-REP offers support for multiple programming languages (see table 5) so that the geometric objects can be accessed remotely. However, it is not possible to directly manipulate the V-REP physics model from SysML because the available API libraries of V-REP do not offer direct support for SysML. Therefore, an intermediate plugin or script is needed that can translate the intended functionality required by the SysML behaviour models into direct object manipulations in V-REP and can also provide feedback. For instance, if the SysML behaviour model requests a forward movement of the vacuum cleaner model, the intermediate script must implement it by manipulating the joint parameter, rotations, positions etc. in V-REP. This plugin is written in the form of python based script and serves to achieve bidirectional exchange of information between

SysML and V-REP. This plugin is referred to as *"python control plugin"* from here onwards.

At this point, it is important to revisit the figure 4-17/figure 4-27 containing the action "*vrep_send*". The SysML behaviour model of the vacuum cleaner first evaluates the input from the interaction device. This input may contain an intent of the VR-user to move or turn the vacuum cleaner or to change the angle of its handle. However, this may cause a collision between the vacuum cleaner and an environmental object that can be detected with the help of V-REP physics model. Therefore, the SysML behaviour model of the vacuum cleaner after evaluating the input from the interaction device sends a command to the python control plugin. This command contains two parameters i.e. [intended movement type, intended handle angle]. An elaboration of these commands can be seen in table 6. The speed of movement of the vacuum cleaner and the speed of change of its handle angle can be fed as an additional parameter with these commands to achieve a more precise implementation.

Table 6: Movement commands sent from SysML to V-REP (python control plugin)

Command	Purpose
[forward, handle++ or handle-- or handle]	Move the vacuum cleaner forward and increase, decrease or don't change its handle angle
[backwards, handle++ or handle-- or handle]	move the vacuum cleaner backwards and increase, decrease or don't change its handle angle
[turn right, handle++ or handle-- or handle]	turn the vacuum cleaner right and increase, decrease or don't change its handle angle
[turn left, handle++ or handle-- or handle]	turn the vacuum cleaner left and increase, decrease or don't change its handle angle
[no movement, handle++ or handle-- or handle]	do not move the vacuum cleaner and increase, decrease or don't change its handle angle

As table 6 mentions, the movement can have five possible values i.e. move forward, move backwards, turn right, turn left and do nothing. The handle angle values just indicate a positive or negative change or no change in the vacuum cleaner handle's inclination. One of these commands is sent out of the SysML behaviour model using "*vrep_send*" action (see figure 4-17) over a TCP/IP connection during each execution. This connection points to the python control plugin that receives the values and the basic functionality of this plugin can be seen as a graphical representation in figure 5-10.

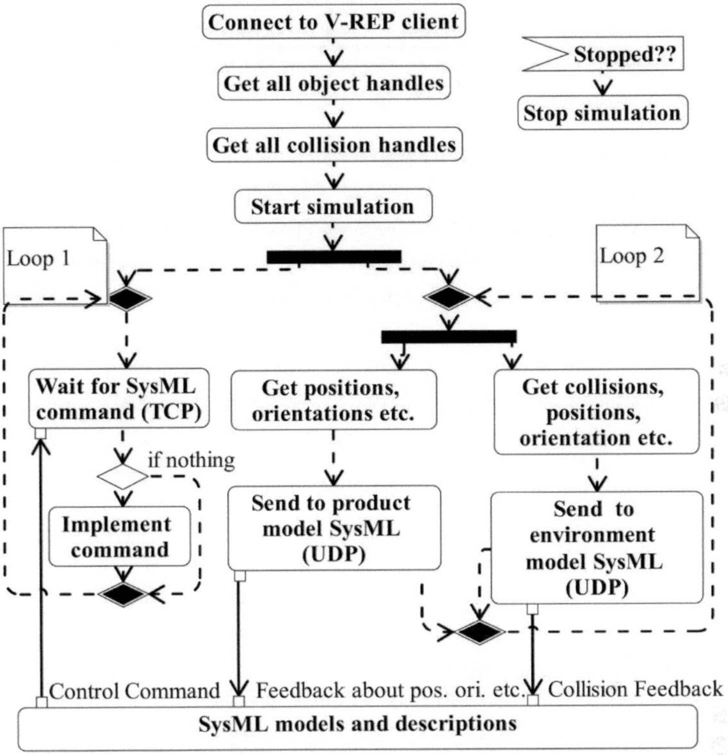

Figure 5-10: Python plugin for V-REP model control

The plugin connects to the V-REP client at the start and acquires the handles for the objects of interest in V-REP. The object handle[65] is a unique identifier for an object that is present inside the V-REP model. For instance, each joint in the vacuum cleaner model has its own identification handle so that each joint can be later accessed by the python control plugin. Similar to the objects, there are collision handles that refer to the collision event between two certain objects. For instance, the collision of the vacuum cleaner with the sofa and with the table will have

[65] V-REP object handle: Object handle is a programmatic term that serves as a unique identifier for a given object in V-REP model

unique identifiers. The collision detection is implemented here by using the default collision detection module inside V-REP. As mentioned in table 5, V-REP offers multiple physics engines. One of these physics engines can be chosen by using GUI or employing an API command. After acquiring all the object handles and collision handles the simulation is started and the physics calculation module against the chosen physics engine is now active.

Once the simulation is active, two parallel programming threads (indicated as *Loop 1* and *Loop 2* in figure 5-10) are launched. The first loop (*Loop 1*) keeps a check on the connection with SysML behaviour models and as soon as a movement command (see table 6) comes in, it is implemented. The *"Implement command"* interprets the SysML command and manipulates the wheels of the vacuum cleaner model in V-REP. For instance, the forward movement is achieved by assigning the wheels with rotation speeds and rotational force values that results in a forward movement of the vacuum cleaner model. Furthermore, the SysML command regarding the vacuum cleaner's handle is implemented by calculating and changing the value of the joint that is present in between the vacuum cleaner handle and the vacuum cleaner base.

The second loop (*Loop 2*) is a feedback loop that constantly delivers feedback about the positions, orientations etc. of objects as well as collision statuses using two UDP connections. The reason for using UDP connection here is the fact that the values can be continuously sent out to a certain UDP port even in the absence of a UDP listener. This means that the feedback is continuously provided even if the SysML behaviour models are not listening. As soon as SysML requires feedback, updated feedback is readily available for use. *Get position, orientation etc.* refers to the feedback about positions and orientation of the vacuum cleaner model in V-REP. This feedback is received in SysML using *rec_pos_Vrep* action in figure 4-17. *Get collisions, position, orientations etc.* refer to the objects in V-REP model that correspond to the environment model. Visually, a collision is indicated inside V-REP by showing the colliding objects in orange colour. In programming, a collision is indicated by a collision flag of value 1 that can be read using the collision handles. This means that a collision flag of value 1 indicated that the two objects defined in the corresponding collision handle are colliding with each other. This feedback is received by the living room model in SysML using "*rec_Simu*" action from figure 4-19. Based on this feedback, the living room model in SysML can determine which two objects are currently taking part in a collision. The action *vrep_send* and *rec_pos_Vrep* from figure 4-17 along with *rec_Simu* from figure 4-19 are "opaque actions" in SysML that contain a chunk of object-oriented code to implement as TCP/IP or UDP based socket connections (see *Annexure A*).

In this way, the python control plugin mediates between SysML behaviour models and V-REP, thus it facilitates

- direct control of V-REP physics model from SysML behaviour models,
- collection of real-time feedback about objects present inside V-REP model,
- provision of this feedback back to SysML behaviour models and
- eventually, two-way communication between SysML and V-REP.

The python control plugin also helps to keep the SysML behaviour models free from detailed implementation information about V-REP physics model. As a result, the modelling effort in SysML can also be spared. In this way, the V-REP model can directly be controlled form SysML behaviour descriptions. This adds to the capability of SysML behaviour descriptions as they now also incorporate indirect physics calculations with the help of V-REP. The next step is to integrate the SysML behaviour descriptions with VR-software so that an interactive simulation of the vacuum cleaner inside the living room can be obtained in VR. This integration is discussed in the next section.

5.3 Integration of SysML behavioural descriptions with VR

After achieving a bidirectional communication between SysML behaviour descriptions and V-REP in the last section, the SysML behaviour descriptions contain the updated properties of virtual objects in real-time. Now, this section is going to explain

- the integration of VR interaction device with SysML and
- the transfer of the updated objects' properties from SysML to VR-software.

The interaction device for the CAVE type VR-system available at Technische Universität Ilmenau (see also figure 2-9) is based on an ART[66] tracking system. This tracking system supports flystick as interaction device and its data can be read in

[66] ART – Advanced Real-time Tracking: https://ar-tracking.com/ [last accessed on 09.03.2020]

real-time. The flystick data can be read in multiple predefined string formats. *6df format*[67] of the flystick is used in this thesis and is detailed in table 7.

Table 7: Flystick data string format [Art17]

6df Format: 6df 1 [id qu bt][s_x s_y s_z η θ φ][b_0 b_1 b_2 b_3 b_4 b_5 b_6 b_7 b_8]	
id	identification number of interaction device (starting with 0)
qu	quality of tracking (-1.000 to 1.000)
bt	joystick value (0, 16, 32, 64 or 128 refer to up, down, left, right and no movements)
s_i	position of flystick (x, y, z values in millimetres)
(η θ φ)	orientation angles of flystick
b_i	the orientation of flystick as a rotation matrix
Example string: 6df 1 [0 1.000 64][261.103 116.520 41.085 19.6522 -57.3530 116.5992][-0.241543 0.968868 -0.054332 -0.482366 -0.168461 -0.859619 -0.842010 -0.181427 0.508039]	

To achieve an intuitive interaction, the height of the flystick from the floor is taken as a measure to calculate the angle of vacuum cleaner handle. *bt* represents the multifunction joystick button and is used as input for movement of the vacuum cleaner (forward, backwards, left, right or no movement). The flystick tracking data is available as a string value over a UDP connection. The interaction device model in SysML (see 4.3.3.3) uses the "*rec_udp*" action (figure 4-21) to receive the flystick data as a string. The received string is evaluated to extract the intended movement and to calculate the angle of vacuum cleaner handle. In this way, the interaction device is integrated with the SysML behaviour models.

The next step is to communicate the updated properties of virtual objects from SysML to VR-software so that the VR-scene contents can be updated. These updated properties mainly involve

- updated positions of virtual objects,
- updated orientations of virtual objects and
- names of objects taking part in a collision.

[67] Pre-defined by ART tracking

The updates about the positions and orientations are transferred to VR-software using the build-in functionality of RTT Deltagen. RTT Deltagen offers direct access to the transforms (position, orientation, scale, centre etc.) of virtual objects present inside the VR-scene using predefined commands (instruction based communication). These external commands can be sent to RTT Deltagen over a predefined TCP/IP connection and once such a commend is received, RTT Deltagen implements it using its default functionality. A few of the commands can be seen in table 8.

Table 8: External commands in RTT Deltagen[68]

Command	**Description**	**Example syntax**
TRANS	Sets the absolute position of an object	TRANS object1 0 10 0
ROT PHR	Sets the absolute rotation of an object in its object coordinate system. PHR = Pitch, Head, Roll	ROT PHR object1 2.2 0.8 0.0
ROT XYZ	Sets the absolute Euler rotations of an object in degrees	ROT XYZ object1 45 60 30
ROTCENTER	Sets the absolute position of the centre of rotation of an object	ROTCENTER object1 2.5 10.9 8.7

TRANS and *ROT PHR* commands are mainly used to communicate the updated positions and orientations values to VR-software form SysML. The action "*VR_Send*" from figure 4-17 is used to construct a string of data consisting of a combination of *TRANS & ROT PHR* commands and is sent to VR-software. As mentioned in figure 5-5, the difference in the coordinates systems is also kept in consideration while forming this data string. As soon as this data string consisting of the updated properties of objects is received by VR-software, the transforms and properties of the virtual object in VR-scene are updated. As a result, the VR-user can observe visual changes in the positions and orientations of objects.

There is no built-in functionality of RTT Deltagen to show a collision indication e.g. a colour change etc. Therefore, the visualisation of collision in VR is implemented by using the API interface of RTT Deltagen. The same idea as present in

[68] Contents in this table are taken from the user manual of RTT Deltagen version 12.2

V-REP i.e. showing the colour of colliding objects in bright orange colour is used to indicate a collision object. This is achieved by a small add-on that is written in the default programming language in RTT Deltagen i.e. C++ and the basic functionality of this add-on can be understood using figure 5-11.

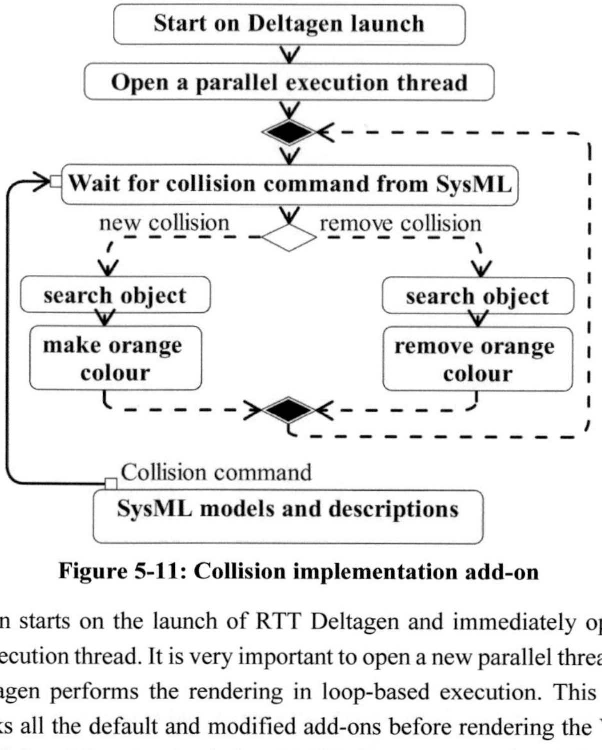

Figure 5-11: Collision implementation add-on

The add-on starts on the launch of RTT Deltagen and immediately open a new parallel execution thread. It is very important to open a new parallel thread because RTT Deltagen performs the rendering in loop-based execution. This execution loop checks all the default and modified add-ons before rendering the VR-scene. If a parallel thread is not opened, then RTT Deltagen gets stuck on *"Wait for collision command from SysML"* and only updates the VR-scene once a value is received from SysML. This can make the rendering of VR-scene contents extremely slow. Therefore, a parallel execution thread is opened so that the waiting time for a collision command from SysML can no longer influence the speed of rendering. After opening the parallel execution thread, the add-on goes into an endless loop. This loop waits for a collision command form SysML. There are two kinds of collision update commands sent from SysML to VR-software and can be seen in table 9. As soon as the collision command is received, the add-on extracts the object name from the collision command. If a new collision object comes in the command, its virtual object is the VR-scene tree is searched, found and changed to bright orange colour. If the command contains the name of the default object[69] (predefined during add-on development), all the collision indications are removed by setting the orange coloured objects back to the default colour.

Table 9: Collision commands sent from SysML to VR-software(CAVE)

Command	Purpose
<mode | obj=objectName>	Make the colour of "objectName" virtual object bright orange
<mode | obj=defaultObject>	Reset all visible orange colour objects to their original colours

In this way, the updated values present inside SysML behaviour models about the positions, orientations and collisions are transferred to VR-software. This enables the SysML behaviour models to directly manipulate objects in VR-software using the commands mentioned in table 8 and table 9. RTT Deltagen independently renders the contents of VR-scene in real-time and together with the information from SysML, provides the user with a product simulation in VR.

5.4 Generic information flow during VR simulation

A CAVE type VR-system is used for example in this chapter to explain the complete development process in detail. However, the VR-model descriptions based on descriptions external to VR-software may also possess the reusability capability across different VR-systems. Figure 5-1 shows the overall flow of information between different tools participating in the VR simulation but is specific to CAVE type VR-system. The same can be extended to develop a more generic information flow diagram as shown in figure 5-12.

[69] An object that is invisible to VR-user and is used for resetting the collision indications. Indicating a collision on this object removes all present collision indications.

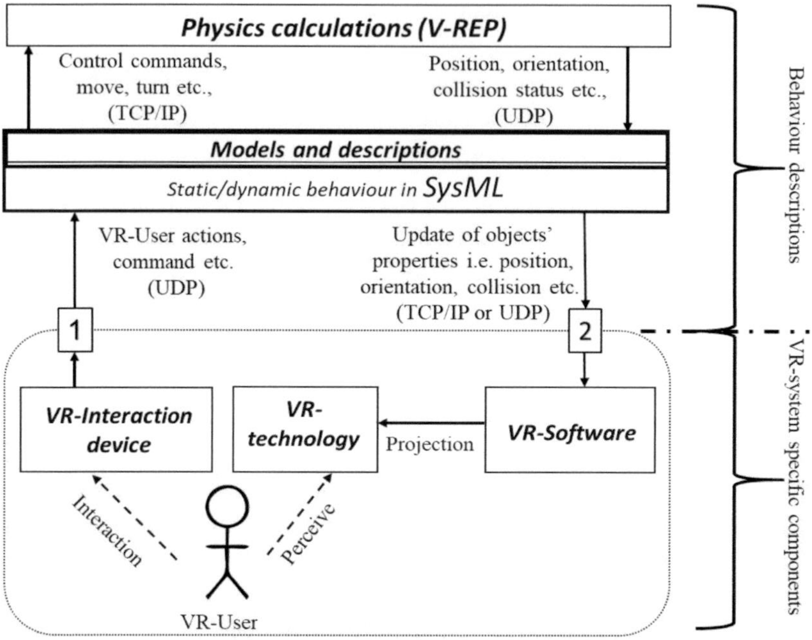

Figure 5-12: Flow of information during a VR simulation (extended from [Mah19a])

The upper half of the figure refers to the behaviour descriptions of VR-model and the lower half refers to the VR side of the story i.e. VR-software, -hardware and -interaction devices. To use the same behaviour description in different VR-systems, the interfaces marked as *1* and *2* in figure 5-12 may require one-time modification against each VR-system. VR technology refers to different VR-systems currently available. Only by managing interface 1 and 2, it is possible to reuse the VR-model descriptions in different VR-systems. In this way, VR-model descriptions based on SysML with the help of V-REP may build a VR technology-independent description method. The next chapter will provide proof of reusability of these behaviour descriptions across different VR-systems by achieving example simulations in an HMD and sVR.

5.5 Summary

This chapter supplements chapter 4 and explains the method for achieving simulation in VR. The development of a visual model of the living room and the vacuum cleaner is discussed in detail. The idea of usage and integration of a dedicated physics calculation software external to VR-system is presented. The development of the physics simulation model and its integration with SysML behaviour models is discussed in detail. Furthermore, the integration of interaction device input from VR in SysML behaviour models and the transfer of updated objects' properties from SysML to VR are discussed in detail. As a result of the presented VR application, it is clear that SysML behaviour models can control the physics calculation on one side and can update the visual model in VR-software on the other side. The overall simulation in VR could be achieved with real-time conditions. A VR-model description performed using SysML and outside of the VR-software also possess the potential to be reused in different VR-systems. Based on the knowledge gained while achieving first example simulation in CAVE type VR-system, a general flow of information involving reuse of VR-model descriptions in different VR-systems is also proposed. The reuse of VR-model descriptions in different VR-systems is further discussed and proof of this concept is provided in chapter 6.

In this way, this chapter has contributed to answering the research question 3 (section 3.5) by deriving a VR simulation with SysML behaviour models. The sub-question II (section 3.5) is partially answered as well by proposing a reuse concept and generic flow during a VR simulation in different VR-systems. Furthermore, the incorporation of real-time physical calculation inside SysML with the help of V-REP has contributed to answering the sub-question III (section 3.5).

6 VR prototypes

Chapter 4 has put forth the method for the description of VR-models using SysML behaviour models followed by chapter 5 that demonstrated the use of these SysML behaviour models to achieve a product use case simulation in VR. This chapter takes a deeper look at the application of the developed method and the reusability of SysML behaviour descriptions in different VR-systems. In this regard, two VR prototypes i.e. the vacuum cleaner and 6 degrees of freedom (DoF) robot respectively are going to be presented in this chapter. With the help of these VR prototypes, this chapter will contribute to answering research question 3 partially and sub-question III completely (see section 3.5).

6.1 VR prototype 1: Vacuum cleaner

As already mentioned in the former part of this thesis, a vacuum cleaner model is considered as a product inside a living room environment, as a case example to demonstrate the modelling method (chapter 4) as well as the simulation process (chapter 5). In this section, the already developed SysML behaviour models of vacuum cleaner, living room and interaction device along with their physics calculation models in V-REP are reused to achieve VR simulation in different VR-systems.

6.1.1 CAVE

The VR simulation inside CAVE type VR-system is already described in chapter 5. Therefore, it is not described again in this sub-section. The reader can refer to the modelling methodology in chapter 4 and the simulation concept in chapter 5 for a detailed explanation about VR simulation inside the CAVE type VR-system.

6.1.2 HMD

An HMD offers a cost-effective and relatively mobile VR-system that is accessible to small and medium-sized enterprises as well as domestic users. The important aspects related to the hardware setup of an HMD have already been discussed in 2.2.2.2. This sub-section uses HTC Vive as the HMD for achieving a VR simulation. The same SysML behaviour models as used to achieve VR simulation in the CAVE are used here. However, the VR technology-specific components (see figure 5-12) have to be adjusted before a simulation can be achieved inside an HMD. In other words, the lower half of figure 5-12 has to be modified and its HMD specific form can be seen in figure 6-1.

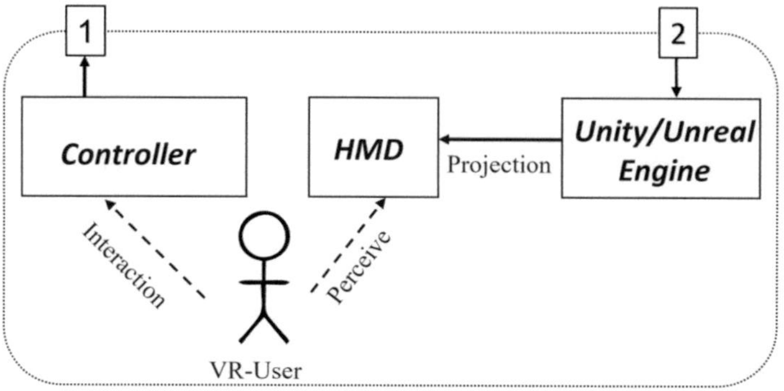

Figure 6-1: Simulation in HMD ([Mah19a])

Two of the most significant VR-software for HMDs are the gaming engines, i.e. Unity3d and Unreal Engine. Unity3d uses C# based scripting to extend its functionality and Unreal Engine uses C++. Unity3d is used in this prototype due to personal preference and past experience with C# based scripting. To achieve the VR simulation in the HMD, the following development-related aspects have to be addressed i.e.:

- Development of a visual model in Unity3d
- Adjustment of interface point 1 to integrate HMD's interaction device with SysML
- Adjustment of interface point 2 to receive the updates of objects' properties in Unity3d from SysML

6.1.2.1 Visual Model

The visual model of the vacuum cleaner and the living room from RTT Deltagen (figure 5-3) is used to construct the visual model inside Unity3d. RTT Deltagen supports the export of complete visual model in FBX[70] format. The FBX format can carry information about

- objects' sizes,
- objects' surfaces,
- objects' transform (positions, orientations etc.),

[70] FBX – Adaptable file format for 3D animation software: https://www.autodesk.com/products/fbx/overview [last accessed on 09.03.2020]

- textures,
- materials
- and object hierarchy etc.

Unity3d offers native support for the import of models in FBX format. The FBX model from RTT Deltagen is imported into Unity3d and as a result, an exact replica containing all the geometric objects, surfaces, materials, textures as well as hierarchy tree is constructed inside Unity3d. During the transfer of the visual model from RTT Deltagen to Unity3d, a few of the objects lost their textures and materials based information. Therefore, the texture and materials are adjusted manually to match those of the visual model of RTT Deltagen.

Hence, the (almost[71]) exact visual model used for the CAVE type VR-system is constructed for use in HMD.

6.1.2.2 Integration of SysML behavioural descriptions with VR

Interaction device integration (interface 1)

Similar to the CAVE type VR-system, HMDs also possess a dedicated tracking system consisting of usually two tracking cameras. These tracking cameras can track the interaction controllers as well as the HMD. The tracking data contains information about the positions, orientations, buttons, triggers as well as the touchpad related real-time data of the controllers. Furthermore, the position and orientation of the HMD itself are also available inside tracking data. The tracking values of the controllers, as well as the HMD, can be read inside Unity3d by using the SteamVR Plugin[72]. This plugin also manages the 3D geometric models of the controllers in VR and offers the possibility to configure controller inputs against different physical buttons that are available on the controllers. Figure 6-2 shows an example controller for an HMD in particular for HTC Vive.

[71] Differences between the colouring, lighting and textures tones are possible.
[72] SteamVR Plugin by VALVE Corporation: https://assetstore.unity.com/packages/tools/integration/steamvr-plugin-32647 [last accessed on 09.03.2020]

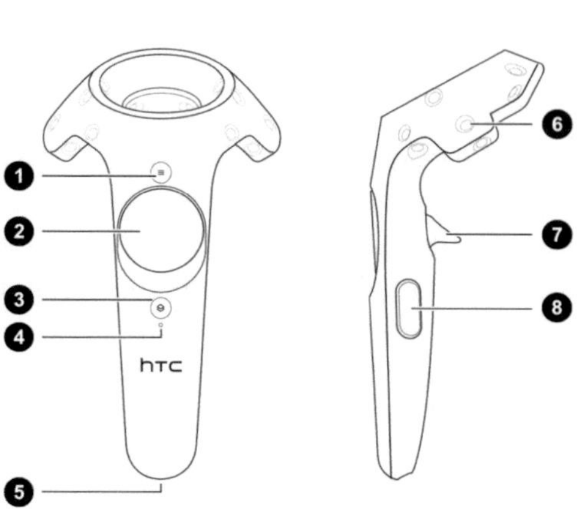

1 - Menu Button
2 - Trackpad
3 - System button
4 - Status light
5 - USB Charger
6 - Tracking sensor
7 - Trigger
8 - Grip button

Figure 6-2: HTC Vive's controller [Htc20]

The tracked position and orientation of the controller is used by the SteamVR Plugin to exactly draw a visual model of this controller inside the VR-scene. The basic interaction is available with the help of the menu button, trackpad, trigger and grip button. For any given application, the feedback from these buttons can be configured to achieve the desired interaction possibilities. To achieve a similar interaction as in the case of CAVE type VR-system, the trackpad is used to control the movements of the vacuum cleaner model and the height of the controller from the ground is taken as a measure to calculate the inclination of the vacuum cleaner's handle. SteamVR Plugin offers the API libraries that allow the access to real-time button values, trackpad values as well as the transform information of the controllers. Unity3d allows by default the scripting in C# programming language. Therefore, a script is written in C# that reads and provides feedback about the controllers' button status, trackpad values, positions, orientations etc. The basic functionality of this script can be understood as shown in figure 6-3. This script is attached to the virtual object corresponding to the controller and starts executing as soon as the scene is played in Unity3d. The scene play refers to the start of VR application in HMD where the virtual objects start to be projected onto the HMD.

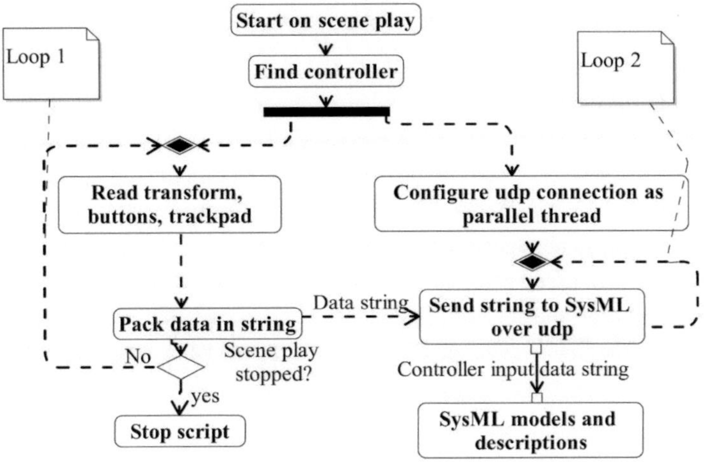

Figure 6-3: Controller script

The controller script searches for the controller of interest and once it is found, a UDP socket connection is opened in a separate parallel thread. This UDP socket refers to the interaction device model in SysML (see 4.3.3.3). The purpose of opening a parallel thread is to have the faster transmission of controller feedback to SysML and to abstain from causing any delay in the main rendering loop of Unity3d. Once the UDP connection is configured, the script moves in a loop (*Loop 2* in figure 6-3) and constantly sends the available data string to *SysML models and descriptions*. Another loop (*Loop 1* in figure 6-3) executes inside the main rendering loop of Unity3d. This loop constantly reads the information about the controller and packs it into a string of the same format as used for CAVE type VR-system i.e. *6df Format* as shown in table 7. The purpose of using the same string format in two different VR-systems is to avoid remodelling of the interaction device model in SysML. Thus, the same format facilitates the use of the same SysML behaviour model of interaction device in HMD as used in the CAVE. In this way, the controller script provides real-time feedback about the controller's data to SysML behaviour models using a UDP socket connection. As a result, the HMD's controller is integrated with SysML and interface 1 (see figure 5-12 & figure 6-1) is configured for use in HMD. The development of this script is typically a one-time task that can be reused in multiple similar applications. The next step is to configure the interface 2 that handles the updated parameters from SysML behaviour models and is discussed next.

Updating virtual objects' properties (interface 2)

As described in section 5.3, the updates about the positions and orientations of objects are handled by the built-in external commands from the RTT Deltagen (VR-software for CAVE). The VR-software for HMD i.e. Unity3d does not offer any such functionality. However, the API libraries of Unity3d can be used to write a C# script that can achieve the same functionality. Furthermore, such a script can also include the implementation of collision visualisation that was handled by an Add-on (see also figure 5-11) in RTT Deltagen. Therefore, a C# script is written to receive objects' updates from SysML behaviour models and to implement them in Unity3d. The functionality of this update script can be understood using figure 6-4.

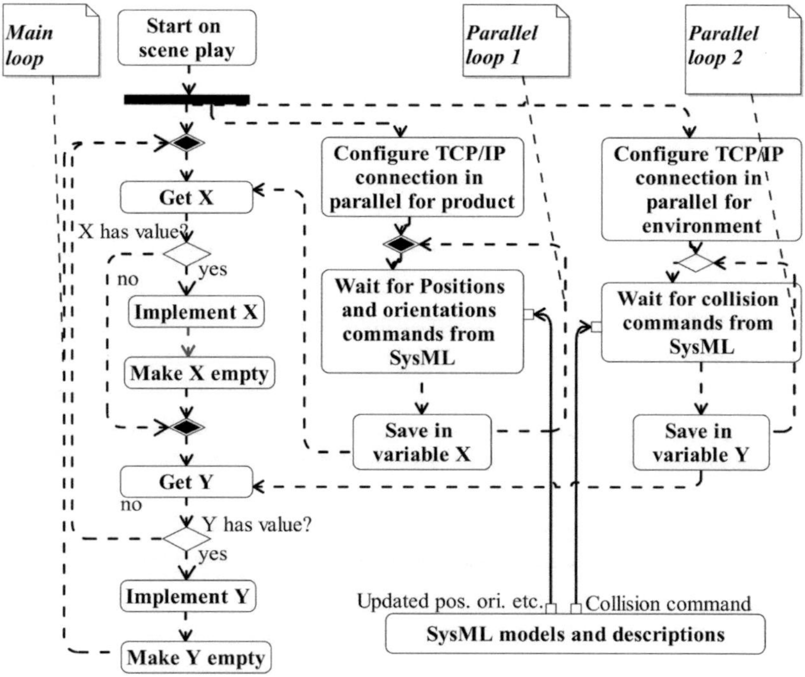

Figure 6-4: Objects' update script[73]

[73] The variable name X & Y are just mentioned for the sake of explanation. The actual variables used inside the implementing script contain other names.

As soon as the scene enters the play mode, two parallel threads are opened and in these threads, two unique TCP/IP receiver socket connections are opened. One connection receives the updates about the positions, orientations etc. from SysML. This thread works as an infinite loop (*Parallel loop 1* in figure 6-4) and *"Wait for position and orientation commands from SysML"* keeps waiting till a command is received. Once a command is received, it saves this command into a variable e.g. *X* in figure 6-4. The second connection receives the updates about collision statuses from SysML. This loop is also an infinite loop (*Parallel loop 2* in figure 6-4) and uses *"Wait for collision commands from SysML"* to receive collision statuses from SysML. As soon as a collision command is received, it is saved in a variable e.g. *Y*. On the left side in figure 6-4, the *Main loop* can be seen that is used to implement these commands. This loop runs in the main rendering loop of Unity3d, however, the implementation is done in a way that it may not cause any unnecessary delay in the rendering. At the start, this loop checks the *X* variable's value to see if there is a command present or if it is empty. If *X* is empty, the execution jumps to check the *Y* variable. If X is not empty and contains a command than this command is implemented. After the implementation variable *X* is made empty. Assigning an empty value to the *X* variable, avoids the multiple times' implementation of a command e.g. in the next rendering cycle, because an empty *X* means that no new values are received from SysML and no implementation is needed. The same logic is used on the value of *Y* and if it contains a command, it is implemented and assigned an empty value. If *Y* is empty, the loop goes back to check the *X* value. This loop runs once during one scene update[74]. *Implement X* and *Implement Y* from figure 6-4 possess a detailed implementation logic that can be seen in figure 6-5.

[74] Unity3d updates a VR-scene multiple times in a second e.g. depending upon the visual contents of the VR-scene the updates may well be around 100 times per seconds.

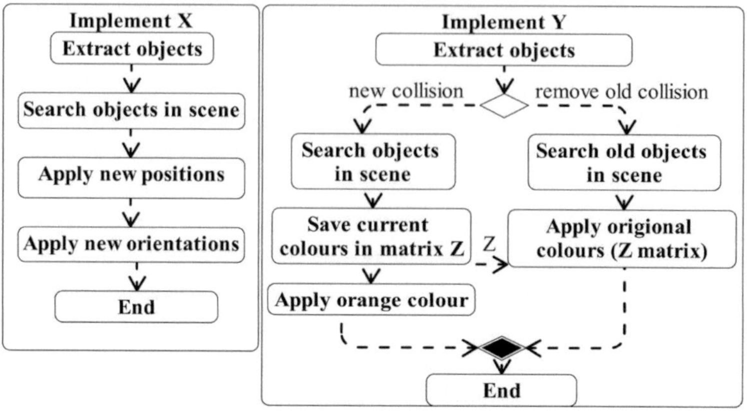

Figure 6-5: Command implementation[75]

Implement X starts by interpreting the received command and extracts the objects mentioned in the command. The extracted objects are searched in the VR-scene and the new position and/or orientation values are applied respectively against the presence of *trans* or *rot* words (see command structure in table 10) in the command respectively. *Implement Y* also interprets the command and extracts the object of interest. At the same time, it differentiates between the *mode* (new collision indication) and *nmode* (remove old collision indication). If the commands request a new collision indication, then before indicating this collision, all the old colours and textures corresponding to the collision object(s) are saved in a matrix *Z*. These colours have to be saved so that they can be applied back to the corresponding object when the collision indication is no longer needed. After that, the colour of the collision object and all its child objects are made bright orange. Hence, a collision is indicted in HMD. Later, if the collision indication has to be removed then it is achieved by applying the original colours and textures previously saved inside matrix *Z*. In this way, the indication and removal of the collision are handled inside Unity3d.

The structure of the commands (position, orientation, collision etc.) originating from SysML is kept similar to that of the ones used in CAVE type VR-system with the slight modification of syntax. The structure of the used commands can be seen

[75] The matrix Z is just mentioned for the sake of explanation. The actual matrix used inside the implementing script contains another name.

in table 10. These slight modifications are made to achieve an easier implementation of these commands in Unity3d and to achieve finer implementation at the same time. For instance, the inclusion of *nmode* command offers the possibility to remove the collision indication from those objects that were previously indicating a collision. This command also eliminates the need for defining a predefined reset object for collision as done in case of CAVE (see table 9).

Table 10: Commands from SysML to VR-software (HMD)

Type	Command syntax	Purpose
Translation	<trans | obj=name | x= value | y= value | z=value>	Translate the object to the given position (x, y, z)
Orientation	<rot | obj=name | rx= value | ry= value | rz=value>	Set the object's orientation. rx, ry, rz are euler angles
Show collision	<mode | obj=name>	Indicate object as colliding
Remove collision	<nmode | obj=name>	Remove the indicated collision

Hence, with the help of the objects' update script the interface 2 (figure 6-1) is managed and SysML behaviour models are integrated with Unity3d for an HMD. Similar to interface 1, the development of interface 2 for HMD is also a one-time task and can be reused for future applications.

6.1.2.3 Summary

The sub-section 6.1.2 has presented the development of a VR-model for HMD that uses the same SysML behaviour models as used in CAVE to achieve a product's use case simulation. By managing the interface point 1 and 2, the already described SysML behaviour models can be reused. The development of the visual model and the scripting methodology to achieve the required functionality are discussed in detail.

There are only two small changes required in the SysML behaviour models because of the change in the command syntax i.e. the script in "*vrep_send*" figure 4-17 and *rec_Simu* from figure 4-19 have to be slightly modified to match the new syntax. Other than these changes, the SysML behaviour models along with the physics model in V-REP can be used as they are used in the CAVE.

6.1.3 sVR

An overview of Smartphone VR (sVR) has already been provided in 2.2.2.3. Smartphones have gained considerable attention recently for the use as VR and AR devices. Small and relatively computationally non-intensive VR games and applications have already surfaced. However, the use of sVR in the product development process is limited which is mainly due to:

- the absence of direct positional tracking and
- the perceptions of sVR as not being a high-performance device deems it non-suitable for industrial application.

The computational and graphical performance of smartphones is improving with every passing day. The highly competitive smartphone market along with increasing expectations/demands of users are pushing the smartphone manufacturer to increase the performance. Almost all the high-end smartphones from all manufacturers that are currently available in the market are capable of executing VR- and AR-based applications. However, the absence of direct positional tracking is a major factor that limits the use of sVR mainly to the visualisation of 360° videos for example [Bor17].

To use sVR for industrial applications, it is very important to achieve positional tracking of the VR-user. The head tracking (orientation only) of the VR-user is already achievable in sVR using Google's VR SDK[76] that works on the feedback from the gyroscope sensor of the phone. However, an application based on the tracking of the orientation of the VR-user's head only allows the user to look around. However, it is not possible to move inside the VR-scene as direct positional tracking against the VR-user's movement is absent. There are already efforts that have tried to achieve positional tracking in sVR e.g. [Fan17]. However, such tracking is achieved by an additional tracking sensor. Along with positional tracking, the quality of visualisation and computational capability of a smartphone are also the topics of concern that are not discussed here.

[Mah19c] points out that for sVR to be suitable for industrial use, the sVR applications shall ensure:

---

[76] Google VR SDK for Android: https://developers.google.com/vr/develop/android/get-started [last accessed on 09.03.2020]

- *a real-time visualisation as well as the simulation,*
- *good quality of visualisation,*
- *reliable position tracking to enable user movements in the virtual space,*
- *easy preparation of VR-scenes and*
- *good interaction possibilities (e.g. integration of an interaction device).*

Therefore, in the light of the above discussion, the capability of a smartphone for use in VR for product development is tested in this sub-section and an implementation idea for achieving real-time positional tracking is presented. Furthermore, the achieved tracking results are analysed quantitatively as well as using a user survey.

6.1.3.1 Visual model

Unity3d offers direct support for smartphone development (both iOS[77] and Android[78]) and therefore, the visual model created in 6.1.2.1 for HMD can directly be used inside the smartphone. The Android development support of Unity3d is used here because the smartphone used in this thesis is Android-based. The main difference between HMD and sVR is that the contents of a VR-scene for HMDs during execution are saved on a computer and are projected live onto HMD, whereas in sVR the complete contents of the VR-scene are compiled and saved in the form of a smartphone application. Such an application may work only using a smartphone and a dedicated computer is no longer needed as is the case with HMDs.

A VR application in the smartphone can be achieved by using the Google VR SDK. Therefore, the libraries for Google VR SDK (v1.150.0) are imported into the visual model for HMD (6.1.2.1) and unity development support is changed to Android. The Google VR SDK divides the smartphone screen in half and creates two distinct imageries that are slightly offset from each other. Using any VR headset/holder for the smartphone holder, for instance, Google Cardboard (also see figure 2-11) is the simplest smartphone holder that makes sure each eye of the VR-user sees a different image and unnecessary environmental light is stopped from entering the eyes.

[77] Apple iOS: https://www.apple.com/de/ios/ios-13/ [last accessed on 09.03.2020]
[78] Android by Google : https://www.android.com/ [last accessed on 09.03.2020]

After configuring the Google VR SDK inside the visual model, the complete contents of the VR-scene can be compiled as an Android application. A screenshot of this application can be seen in figure 6-6.

This application can map the head movements of the VR-user using feedback from the gyroscope sensor inside the smartphone with the help of the functionality of Google VR SDK. As a result, the VR-user can look around (360°) and visualise the contents of the VR-scene. However, this application so far does not contain any positional tracking implementation. An idea about implementing positional tracking without the need for any dedicated tracking sensors is presented next in 6.1.3.2.

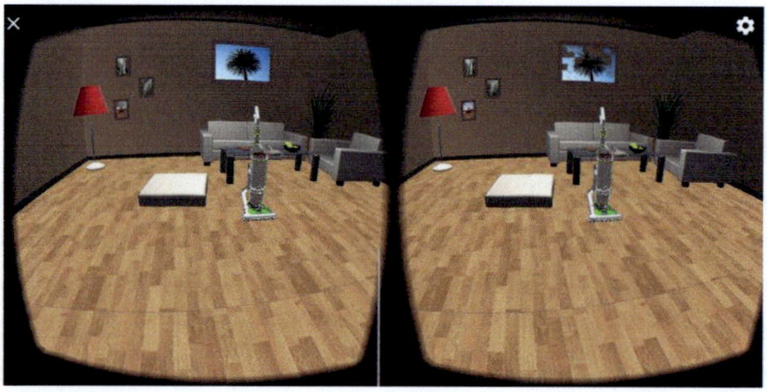

Figure 6-6: VR-scene contents in a smartphone application [Mah19c]

6.1.3.2 sVR positional tracking

The Google ARCore[79] includes positional tracking as built-in functionality. ARCore renders the camera feed of the smartphone directly onto its screen. Virtual objects can be overplayed onto the camera feed and manipulated. The positional tracking in ARCore is based on the environmental understanding with the help of the smartphone camera. The flat surfaces in the surroundings can be identified as references and the movement of the smartphone compared to these references help to provide an estimate of the positional movements. The orientation of the smartphone is tracked by evaluating the feedback from the gyroscope sensor inside the smartphone. As a result, positional tracking is achieved using ARCore in an

[79] Google ARCore for Android: https://developers.google.com/ar/discover [last accessed on 09.03.2020]

AR application. One possible way to achieve positional tracking in sVR is to include ARCore's functionality inside the VR application. Cardboard requires a smartphone with Android 4.4 or higher with gyroscope whereas ARCore requires a smartphone with Android 7.0 or higher with camera and gyroscope sensor. Therefore, a smartphone running either Android 7.0 or a higher version with a camera sensor and a gyroscope sensor is needed as sVR device. Normally, all the High-End smartphones currently available in the market can fulfil these requirements.

The smartphone used in the development and to generate the results presented here possessed the following technical specifications:

- CPU: 8 core processor (4*2.36 GHz + 4*1.8 GHz)
- GPU: ARM Mali-G72 MP12 (12 cores with 850 Mhz each)
- 6 GB RAM
- 407 dots per inch (dpi) with 2240x1080 resolution
- Android 9.0
- Gyroscope and accelerometer

Tracking algorithm

[Men17] has used a similar fusion of ARCore and VR SDK to achieve positional tracking in sVR. However, this work applies a scaling factor of 6 in Y-axis and 10 in both X- and Z-axis values after reading the tracked position values from ARCore. If such scaling factors are applied, the achieved positional tracking cannot be of comparable proportion to the physical world movements. Although the experiments conducted in the scope of this thesis use a similar idea, the mapping of 1:1 is used and no scaling factor is applied. The purpose of 1:1 mapping between ARCore positional values and the positional movement of the virtual camera in the VR-scene is to achieve positional tracking comparable to the VR-user's physical movements.

The visual model with already loaded VR SDK libraries is extended to include ARCore (v1.5.0) libraries for Unity3d. ARCore libraries tend to render the camera feed onto the smartphone screen. As in a VR application, the camera feed is not required, it is stopped and only the content rendered by the VR SDK is left for VR-user's view. To integrate the positional tracking capability of ARCore in the sVR application a small position tracking script is written and attached to the camera object in the scene. The functionality of this script can be understood by figure 6-7.

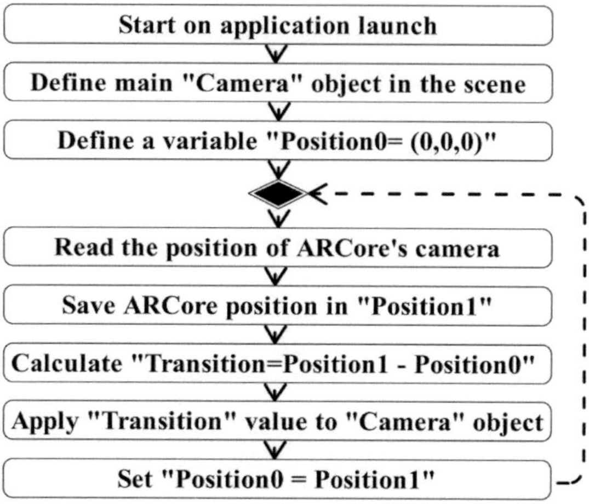

Figure 6-7: Positional tracking script

Once the smartphone application containing the positional tracking script is launched, the script defines a "*Camera*" object that refers to the main camera[80] in the VR-scene. A new variable *Position0* is defined and assigned absolute global origin (0,0,0). From here onwards, the script enters into an endless loop, that reads the values of ARCore's camera and saves it into *Position1* variable. *Position1* refers to the new position of the VR-user and a comparison of it with the old reference value i.e. *Position0* provides the relative change in the position. This relative change is saved inside the *Transition* variable and is applied to the *Camera* object. In this way, the positional movements of the VR-user from real-world are translated to the positional change of the virtual camera in the VR-scene. As a result, the direct positional tracking in sVR is achieved with the help of ARCore and the smartphone's camera. In the end, the new position (*Position1*) is saved in Position0 for use in the next execution cycle.

[80] The position of main camera decides what content will be shown to the VR-user or in other words, it is the view point in the VR-scene

Accuracy of sVR's positional tracking

A working positional tracking is not enough to justify the use of sVR as a device for VR in industrial applications. Therefore, a judgment about the preciseness and quality of the achieved tracking shall also be made. There can be two possibilities to measure the quality of positional tracking i.e.

- manually by moving smartphone between measurable marked points or
- by comparing it with an HMD containing a dedicated tracking system.

The manual measurement can be prone to error in multiple ways as the global origin (0,0,0) in ARCore refers to the position of the smartphone's physical camera. The camera position can be different on different smartphones based on their design. Furthermore, keeping the width and thickness of a smartphone in mind, it is extremely difficult to locate the exact position of the physical camera that refers to (0,0,0) position. As a result, an unmeasurable offset can be induced in the measured values. The second possibility is to compare the positional tracking in sVR with that of an HMD. An HMD usually has two tracking cameras and works on marker-based tracking principle to track the position as well orientation of the HMD and controllers. This approach is adopted to achieve a measure of the quality of sVR's positional tracking. There are three main reasons for choosing such a comparative method because

- such an experiment is easy to set-up,
- it is less error-prone as compared to the manual method and
- it is easily reproducible.

The physical construction for this experiment involves the mounting of the smartphone directly onto the HMD as shown in figure 6-8. Caution must be exercised in the understanding of this experiment and the results that follow i.e. the HTC Vive's tracking is assumed to be perfectly accurate and a reference for the comparison. Furthermore, it should be made clear that the author does not make any claim about the quality of the tracking of HTC Vive.

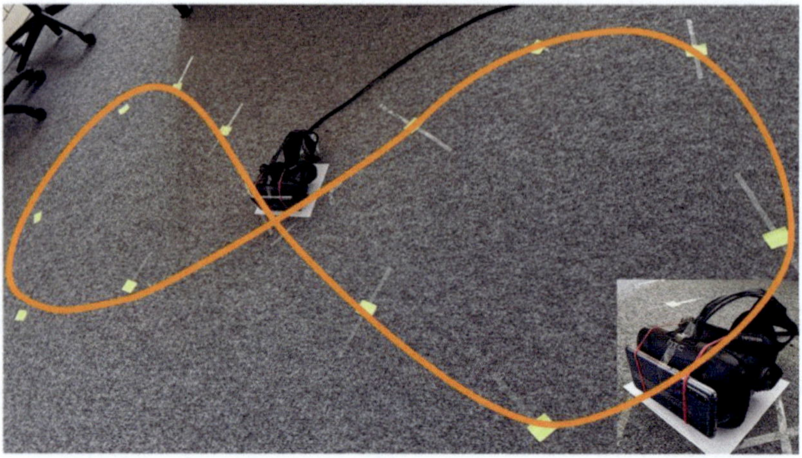

Figure 6-8: Physical construction

Several reference positions within the tracking range of HTC Vive are marked on the floor to form a shape of the digit "8". Two empty Unity3d VR projects are constructed that contain no visual objects and can only store the tracking values for both HMD and sVR respectively. After starting the VR application on the smartphone, it is mounted onto the HMD and placed on the start position (big white spot in the middle). Now, exactly one value each containing the 3D coordinates of the Vive's and sVR's tracking values are received and saved in separate text files. The same procedure is adopted to record the values for all the marked points while completing a movement along the shape corresponding to the digit "8". In performing this manoeuvre, efforts are made to avoid any drastic change in the orientation of both the devices so that a clean comparison can be obtained. Finally, a total of 15 values are recorded in individual text files. The first tracking values for both devices are shifted to (0,0,0) to achieve uniform tracking comparison and the remaining tracked values are also shifted accordingly. As during the experiment both devices are always placed on the floor, the height value is neglected for this comparison and the resulting 2D plot can be seen in figure 6-9.

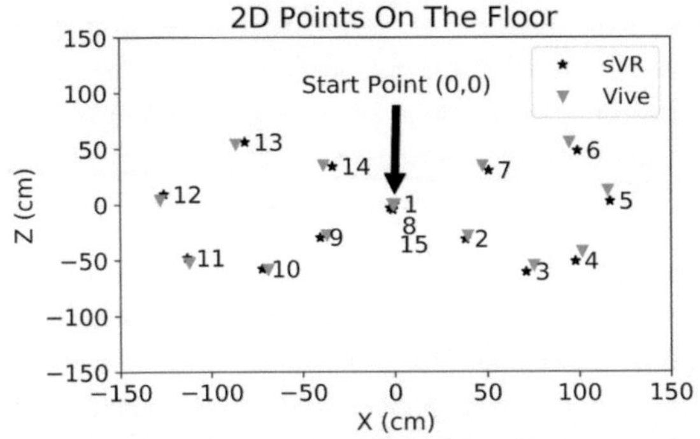

Figure 6-9: Tracking values from both devices (i.e. sVR & Vive)

The numbers 1 to 15 placed next to the measured point indicate the sequence in which the tracking values are recorded. sVR's tracking followed the path of the Vive's tracking with some deviation. Although both devices are always placed on the floor before recording a value, there are still variations[81] in the y-axis values as well. In the case of Vive, the y-axis values vary between -4.4cm to1.4cm and for sVR, it is between -4.1cm to 0.5cm. To calculate the absolute positional error (i.e. the distance between both tracked points), all three axes (x, y, z) values are considered.

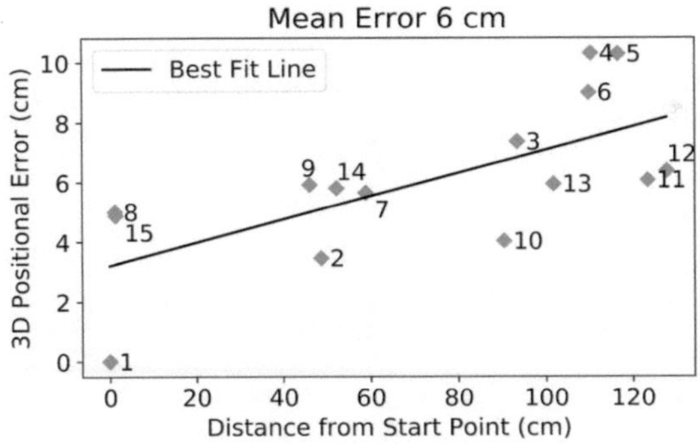

Figure 6-10: Positional error

[81] One possible cause of these deviation can be an inclined placement of HMD on the floor that causes the HTC Vive's markers to have different height at different calculation points.

Figure 6-10 shows the distance between tracked values from both devices for a given marked point on the floor as the positional error. The error increases (approximately) linearly as the distance from the starting point increases as can be seen by the best fit line plotted on figure 6-10. The average error for the 15 measurements done in a 2D movement space of approximately 2.6x1.2 meters is 6 cm.

It can be argued that a 2D measurement can only deliver a vague comparison about the tracking quality as the actual use of sVR tracking is intended for a 3D space. Therefore, a quantitative comparison of points in 3D space is also required which can be achieved by moving both devices in 3D space. This requires the values of the tracking to be recorded continuously. The continuous recording of tracking values can be challenging as both the application are running on different frames per seconds (fps). The smartphone achieves 60 fps while Vive is running on 105 fps. The difference in fps can make a comparison of the recorded values difficult and potentially inaccurate. To have the same number of tracking values at the same instance in time, a delay of 100 milliseconds is used as a workaround i.e. after 100 milliseconds both tracking values are recorded instantaneously. The recorded points in the 3D space are visualised in figure 6-11.

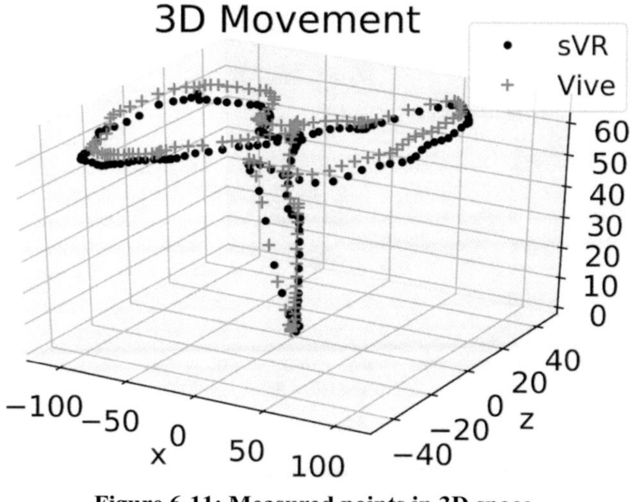

Figure 6-11: Measured points in 3D space

The movement involved lifting both devices that are mounted together, performing a movement that resembles the digit “8” and finally, placing both the devices on the floor. The size of the 3D movement space is approximately 1.0 X 2.4 X 0.7 meters. The positional error between the tracked values from both the devices can be seen in figure 6-12. To have a fair mean error calculation, the tracked values at the start and end of the movement are ignored because both devices are placed on the floor. As a result, a mean error of 5.57 cm is achieved.

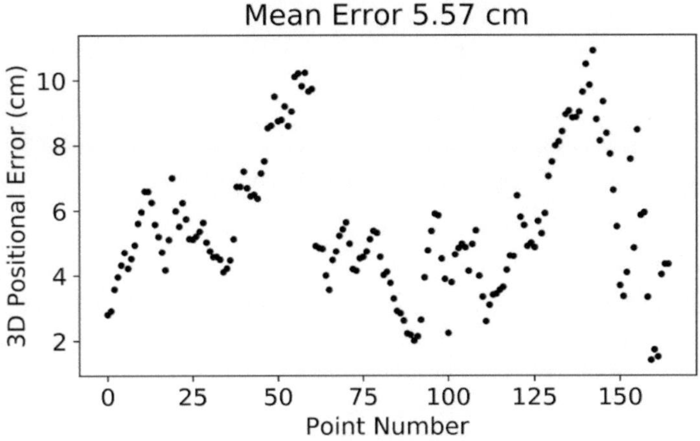

Figure 6-12: 3D positional error

Hence, the conducted experiment has shown that although sVR's positional tracking is not as accurate as of Vive, it lies in a comparable range. After presenting the tracking implementation idea and the quantitative comparison of tracking quality, the integration of SysML behaviour models with sVR is discussed next.

6.1.3.3 Integration of SysML behavioural descriptions with VR

Interaction device integration (interface 1)

After presenting the implementation of positional tracking in sVR, the integration of SysML behaviour models is discussed here. Similar to the case of HMD, the technology-specific components mentioned in figure 5-12 can be adjusted for the sVR and resulting lower half of this diagram can be seen in figure 6-13.

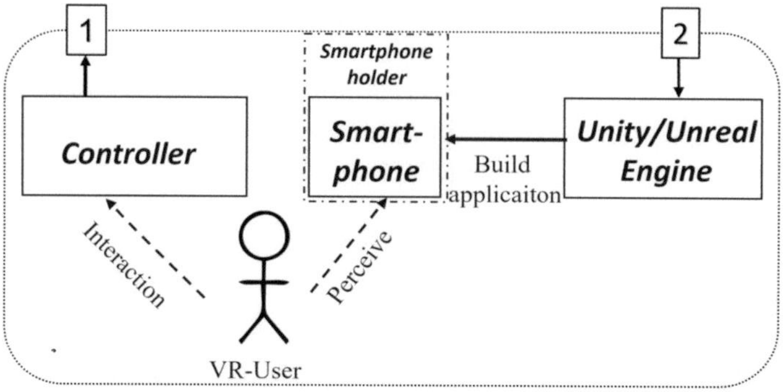

Figure 6-13: Simulation in sVR[82]

In case of CAVE and HMD type VR-systems, the VR-software directly renders the contents of a VR-scene. However, in the case of sVR, it is achieved by first compiling the contents into an executable Android application. This application can be executed with the help of a smartphone's hardware and placing the smartphone inside any available VR glasses/ smartphone holder can complete the setup for a VR device. The controller here can be a Bluetooth gamepad or game controller available for Android devices. Similar to the HMD, Unity3d is used here as the VR-software to develop the smartphone application. Therefore, the plugins developed for HMD are reused for sVR as well with minor modifications. A low cost (<10 €) Bluetooth controller (figure 6-14) for Android devices is used as an interaction device and is integrated into the sVR application.

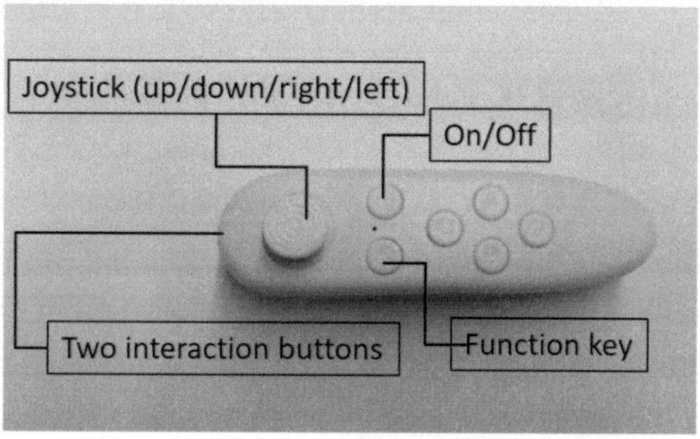

Figure 6-14: Bluetooth controller

82 Smartphone holder: A device to hold the smartphone, possess a construction to block unnecessary light from entering VR-user's eye, let each eye view only half of smartphone's screen e.g. a Google Cardboard

The status of the individual buttons and the joystick of the Bluetooth controller can be read inside Unity3d. These are arranged and packed inside a *6df Format* (see table 7) in the form of a data string. This data string is sent to SysML behaviour models reusing the same controller script (see figure 6-3) as used in the case of HMD. In this way, the Bluetooth controller is integrated with SysML behaviour models.

Updating virtual objects' properties (interface 2)

The same behavioural descriptions in SysML that are already used in CAVE and HMD are reused to achieve simulation in sVR. The execution of SysML behaviour models and the physics engine calculations are executed inside a laptop and updated parameters are transferred wirelessly to the smartphone utilizing a Wi-Fi connection. To receive the updated properties and collision statuses of virtual objects form SysML, the objects' update script (figure 6-4 & figure 6-5) from HMD is reused. It can be difficult at times to establish a reliable TCP/IP socket connection with an Android smartphone because of Android security protocols. Therefore, both the TCP/IP receiver socket connections in figure 6-4 are changed to UDP socket connections. The implementation of the received command is performed in the same way as done in the case of HMD. In this way, SysML behaviour models are integrated with sVR.

Inside an active simulation in sVR, the laptop performs the computationally expensive physical calculation and the execution of SysML behaviour models (see figure 6-15). The smartphone performs the rendering, visualisation and position tracking. Furthermore, the smartphone updates the properties of virtual objects in the scene against the updated objects' properties received from SysML and also provides feedback about the controller state to SysML.

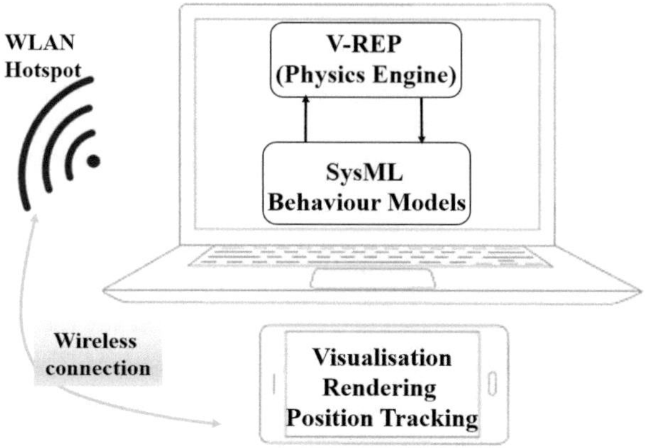

Figure 6-15: Concept of simulation in sVR [Mah19c]

In this way, a real-time VR simulation is achieved inside sVR that uses the same behavioural models as used for CAVE and HMD type VR-systems.

6.1.3.4 Survey

To achieve initial feedback about the use of sVR in product development, a survey was conducted. The primary goal of this survey was to collect first feedback about the sVR's use as a VR-system for industrial application. The *Cognitive Walkthrough*[83] [Lew90] methodology was used to conduct this survey where the test persons were shown two example VR applications. The test persons were given the task to

- familiarise themselves with the virtual environment by walking around,
- access the tracking quality of implemented positional tracking and
- provide potential application area for sVR.

After showing the VR application, the test persons were asked some questions to collect feedback (see the questionnaire in <u>*Annexure B*</u>). The test persons were

[83] A Cognitive Walkthrough is a form of usability study where the users are provided with a system/interface and asked to perform certain tasks. After experiencing the system, the users are asked a series of questions to access the system's learnability by the users.

asked to freely express themselves by filling in self-thought keywords and terminologies to answer the questions in the survey.

The survey included the vacuum cleaner (see figure 6-6) as example 1 that integrated the behaviour models as well. A second example based on the visualisation of an automobile (see figure 6-16) is included in this survey.

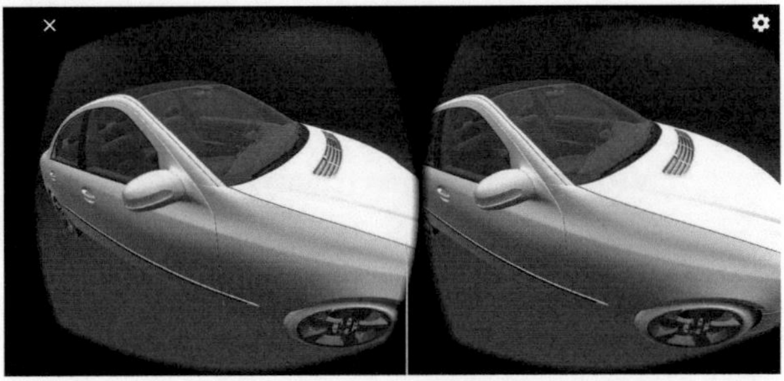

Figure 6-16: Automobile[84] visualisation in sVR [Mah19c]

The control of the vacuum cleaner model is done using the laptop keyboard as at the time of performing this survey, the Bluetooth controller was not integrated inside the smartphone application. The frames per second (fps) for this application varied between 35-60 fps depending on the number of objects that had to be rendered by the smartphone. The second example achieved a frame rate of 40-60 fps at the time of conducting this survey. The purpose of this survey was to simulate industrial design evaluation and visualisation scenarios and was presented to a total of 15 participants. All the participants consisting of 14 males and 1 female had a background in mechanical engineering. Among the participants were professors, researchers, PhD students and 4 alumni members related to a technical university in Germany. The test persons were asked to move inside the virtual space and examine the feasibility of sVR for use in product development. The age distribution of the test persons and their past experience with VR/AR can be seen in figure 6-17.

[84] The geometric model of the car is taken from RTT Deltagen v12.2

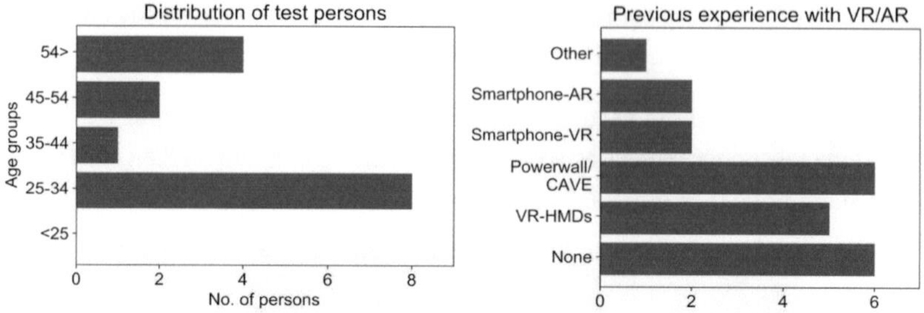

Figure 6-17: Age distribution of test persons (left), past experience with VR/AR- technologies (right) [Mah19c]

The test consisting of both the applications that lasted between 5 to 8 minutes and at the end of the test, the test persons were asked a series of questions. They were asked to mention their expectations from VR technologies and smartphone VR in particular. The key expectations of the test persons from the sVR technology that can be seen in figure 6-18 and from sVR applications that can be seen in figure 6-19.

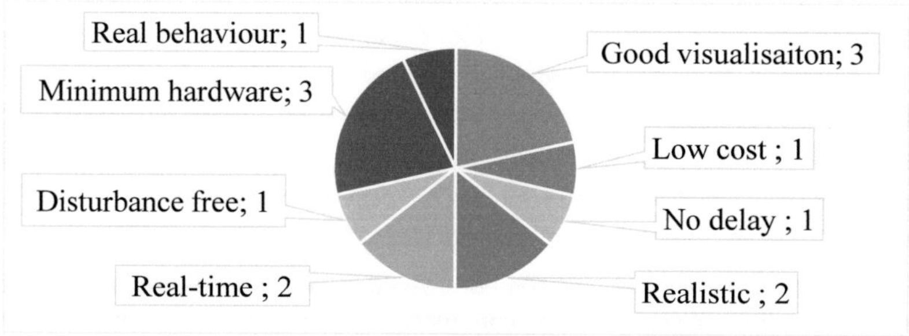

Figure 6-18: What do you expect from sVR technology?

Good visualisation, minimum hardware requirement, realistic visual contents and real-time simulation capability are pointed out by the test persons as key expectations from the VR technologies. In the particular case of sVR, the test persons mentioned the following key expectations:

- Better visualisation
- smoother movements
- easy use
- better interaction

The smartphone used for conducting this survey possessed only 409 dpi that cannot offer a very fine visualisation. This problem can be solved by using a smartphone with higher dpi. The expectation "smoother movements" links to the drop of fps to 35 against higher rendering effort from the smartphone. The reduced fps induce jerks in the positional tracking that induced noticeable steps for the VR-users during the positional tracking and rendering. An increase in the fps can help to overcome this limitation and to achieve smoother movements. Furthermore, the test persons expected an easy and intuitive use of sVR along with good interaction possibilities.

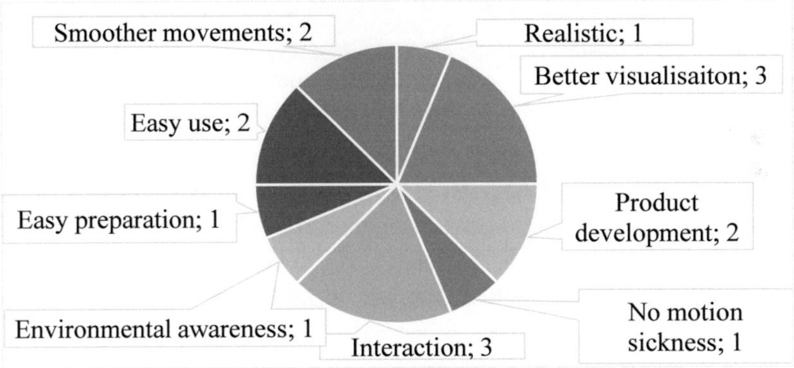

Figure 6-19: What do you expect from sVR applications?

The accuracy of the positional tracking of a smartphone is sensitive to the visibility of virtual markers at all times. If the smartphone camera is covered while conducting the tests, it can lead to inaccurate positional tracking in sVR. The tests were conducted in the proximity of 4m X 8m while all 15 test persons were present within this area. This led to a situation that at times, the field of view of the smartphone was fully covered by the body of a person standing right next to the test person. If the smartphone camera cannot see the virtual markers, the ARCore can lose the track of positional tracking and random jumps start to occur. To overcome this problem, the smartphone application was restarted from time to time. Test persons were expected to rate the quality of tracking. The rating expected the test person to characterise the quality in five levels (very imprecise to very precise) as shown in figure 6-20.

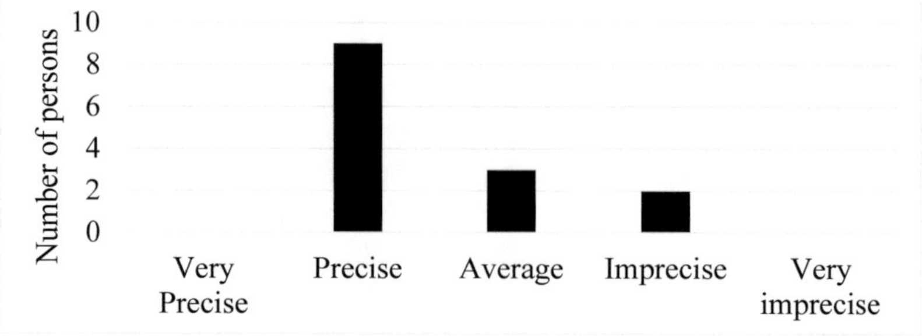

Figure 6-20: The evaluation of tracking quality by test persons

14 out of 15 test persons rated the quality of tracking and the overall rating tends to possess average to precise characteristics. Finally, the test persons were asked to suggest the possible applications of sVR based on its capabilities as depicted by the two presented applications. The test persons were allowed to express freely in the form of keywords and the received feedback can be visualised employing figure 6-21. The use in learning, visualisation, product development and gaming were among the major suggested applications. Under product development, the use in prototyping and the evaluation of Computer Aided Engineering (CAE) was also suggested as application areas by the test persons.

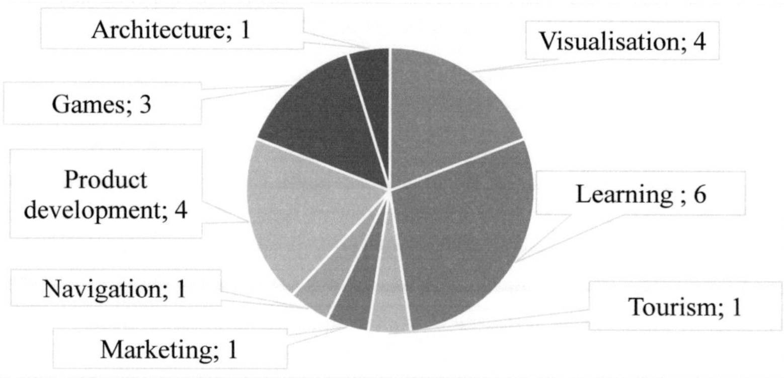

Figure 6-21: Possible application fields for sVR according to test persons

The received feedback against the conducted survey is encouraging as far as the quality of positional tracking and sVR's potential use in industrial applications are concerned. The quality of visualisation can be improved by using a smartphone with higher dpi e.g. smartphones with 540 dpi are already available in the market. Although the sVR tracking is reliable, caution must be exercised so that the field of view of the camera stays free at all times and is not covered. The comparisons of sVR's tracking with HTC Vive shows comparable results, however, sVR is not exactly as accurate as HTC Vive. Therefore, the use of sVR may depend on the level of the preciseness of the tracking needed for any given application.

6.1.3.5 Overall Simulation flow in sVR

Although the overall flow of simulation in sVR (see 6.1.3 & 6.2.2) is based on figure 5-12, a more smartphone specific flow of information can also be developed based on the experiments conducted in sub-section 6.1.3. This information flow specific to the sVR can be seen as shown in figure 6-22.

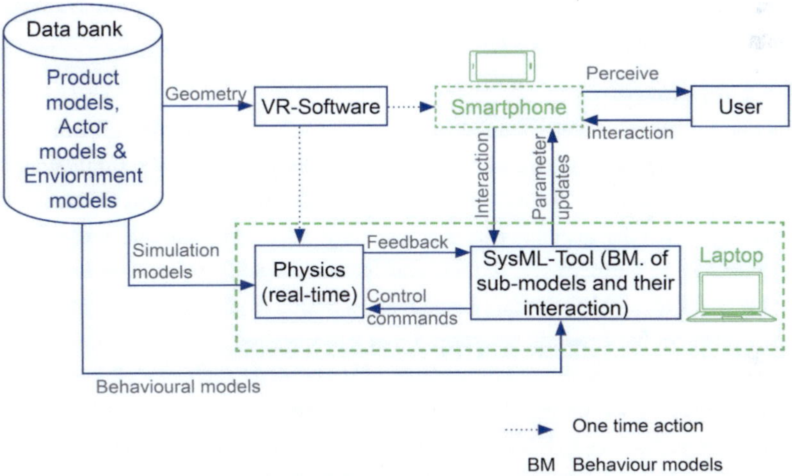

Figure 6-22: Flow of information and execution hardware in sVR (extended from [Weg19])

The geometric models, SysML behaviour descriptions and V-REP model can be loaded inside VR-software, SysML execution tool and V-REP respectively as a one-time process from the model database. The model database contains already developed models and descriptions. The visual content can then be compiled as an Android application from the VR-software. This Android application can be executed stand-alone and it actively exchanges information with SysML behaviour models to achieve an interactive VR simulation. The SysML behaviour models communicate with real-time physics calculations that are performed inside V-REP and updates the content of smartphone application continuously. The interaction with the user is achieved using controller feedback that flows from the smartphone to the SysML behaviour models over a Wi-Fi connection. The smartphone itself performs rendering, visualisation and positional tracking (see 6.1.3.2). The SysML behaviour models and V-REP physical calculations are computationally expensive and therefore, are executed on a high-performance laptop. The smartphone is connected wirelessly with the laptop over a Wi-Fi connection so that information exchange can be made possible.

In this way, sVR builds a cost-effective and mobile VR hardware setup that uses a smartphone, smartphone controller, smartphone holder and a laptop for computation. Such a hardware setup is easy to carry on the way and remote product presentation can also be made possible.

6.1.4 Summary

Section 6.1 has presented a detailed explanation about the use of SysML behaviour models together with physical calculation in V-REP to achieve VR simulations in different VR-systems. The use of the same behaviour descriptions in different VR-systems proves the generality of the behaviour description method presented in chapter 4 & 5. Furthermore, the reuse of the same SysML behaviour description and physics simulation model to achieve simulation in different VR-system partially answers the research question 3 and sub-question III in full capacity (see section 3.5). Thus, the overall preparation effort needed to achieve VR simulations is reduced as the VR preparation for one VR-system can be reused across different VR-systems. As CAVE, HMD and sVR lie in different price budgets, the presented method can target a vast industrial audience ranging from small to medium-sized enterprises up to large enterprises as well.

6.2 VR prototype 2: 6 DoF robot

A vacuum cleaner is typically a consumer product and the presented scenario in this thesis simulates a household product use situation. To simulate an industrial scenario and to show the generic applicability of the developed method, a 6 DoF robot inside an assembling environment is presented in this sub-section. 6 DoF robot as the second prototype demonstrates a 6 DoF robot working on a conveyor belt inside an assembly line. The conveyor belt transports cube-shaped objects that are picked by the robot and placed inside a basket. The robot itself can be operated in automatic and in manual modes. In automatic mode, it continuously picks the objects from the conveyor belt and places them in the basket. In manual mode, the same task can be performed by a user who manipulates the position of the robot's gripper, grips cubes and drops them in the basket by opening the robot's gripper. This scenario is constructed inside HMD and sVR. The construction of VR application for CAVE type-systems is excluded for this prototype. The three sub-models for this VR prototype are 6 DoF robot as the product, a room containing a conveyer belt as environment and interaction device as the actor.

6.2.1 HMD

The same VR equipment (i.e. HTC Vive) and VR-software (i.e. Unity3d) are used for this prototype as they were used for the vacuum cleaner example (see 6.1.2).

6.2.1.1 Visual model

The CAD model[85] of the robot used in this prototype is first exported as VRML format. As Unity3d cannot directly facilitate the import of VRML format, it is converted to FBX format using Blender 2.79. The resulting FBX file can directly be imported into Unity3d and the resulting VR-scene can be seen in figure 6-23.

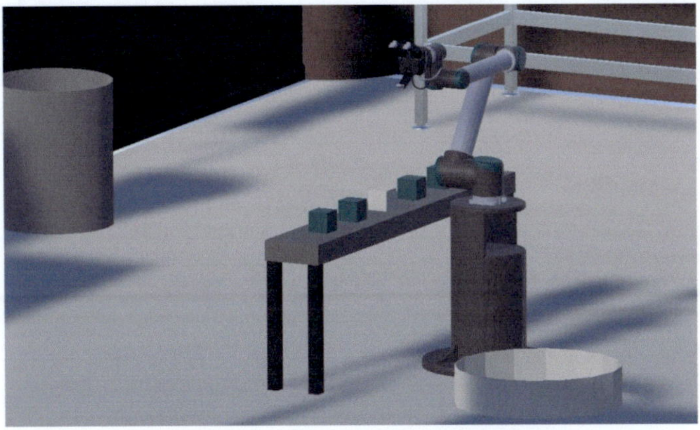

Figure 6-23: Contents of the VR-scene

[85] CAD Model of robot is taken from grabcad.com. URL: https://grabcad.com/library/ur10-robot-w-robotiq-3-finger-gripper-1 [last accessed on 09.03.2020]

The core components of the visual model are the robot model itself, conveyor belt, basket and a simple room model[86]. The room model together with the conveyer belt, cubes and the basket built the environment model for this VR prototype. The same visual environment model is used to build the environment model in V-REP and is explained next.

6.2.1.2 V-REP model

The individual objects present in the room in Unity3d are used to export STL geometric files that are later imported inside V-REP to build the exact replica of the room. The conveyor belt is used from the default objects of V-REP. This conveyor belt is the default asset of V-REP and allows the modification of its size as well as the movement speed. A specific position on one end of the conveyor belt that is in the reach of the robot is chosen as the default position from where the cube can be picked by the robot. The conveyor belt moves till this positions is empty and stops as soon as a cube reaches this place. In this way, the conveyor belt feeds the cubes one after another to the default position so that the robot model can pick them.

The robot model that is available as a CAD model has a total of 6 joints that construct 6 degrees of freedom. It also possesses a gripper with three fingers and each finger has three joints that help it to grip objects. The CAD model is converted to the URDF model on the same lines as already explained in sub-section 5.2.1 that explains the construction of the kinematic model in V-REP. The URDF model is then imported in V-REP and as a result, the completed V-REP model can be seen in figure 6-24.

[86] The room model is taken from free assets at unity Asset Store https://assetstore.unity.com/packages/3d/environments/morgue-room-pbr-65817 [last accessed on 09.03.2020]

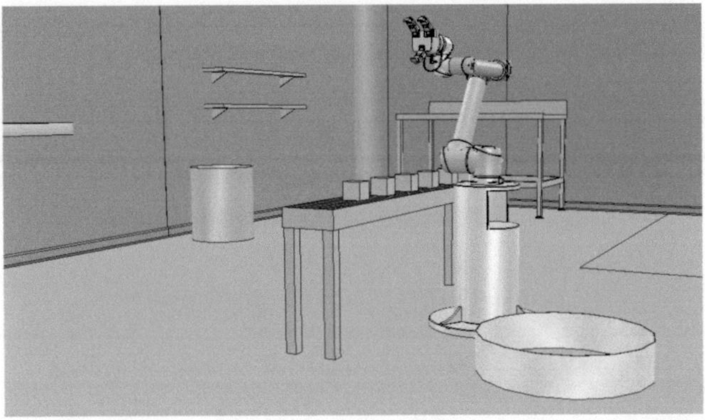

Figure 6-24: V-REP model for the computation of physics calculation

The default collision calculation module in V-REP is used to perform the collision detection between geometric objects. V-REP also offers an inverse kinematic calculation module by default. This inverse kinematic module is used to calculate the exact joint rotations against any given goal position/orientation of the robotic gripper. Furthermore, V-REP also provides pre-implemented examples for path planning for such robots and further similar robotic arms that are used to perform grasping tasks. As a result, the position of the robot gripper can be controlled by instruction commands (e.g. move to a point, set orientation etc.) and automatic path planning can also be used to grip the cubes from the conveyor and drop them into the basket.

6.2.1.3 Development of SysML behaviour models

To control the physics simulation model and to update properties of virtual objects in VR-software, the SysML behaviour models are modelled keeping in view the guidelines mentioned in section 4.4. First, the HLSA model containing the structure of the overall system, its interfaces and signals is modelled. The structure of the overall system can be seen in figure 6-25 and the interconnection of ports can be seen in figure 6-26.

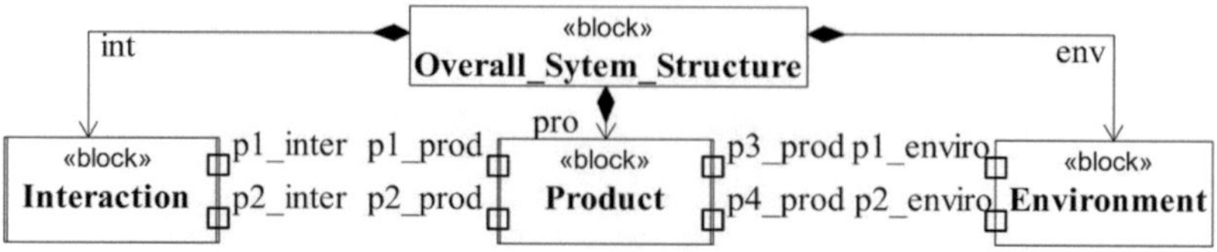

Figure 6-25: HLSA of the complete system

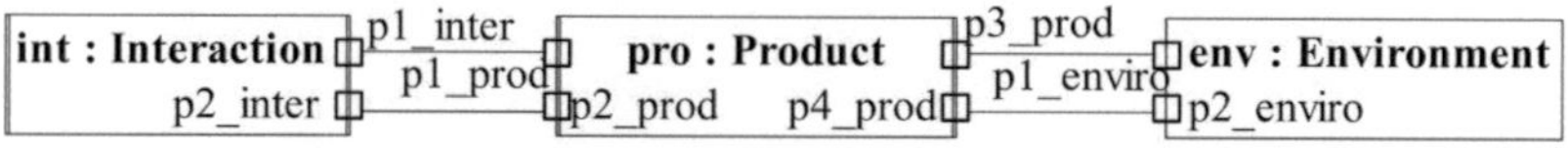

Figure 6-26: Internal structure of the overall system

The structural model of the overall system is saved in a project file and is reused in the modelling of product, interaction and environment models. Furthermore, the possible signals that can flow in between the three sub-models over the defined ports are also modelled and saved in this project file. This project file is loaded and shared inside the product (robot) model that is saved in its own project file. The description of the robot model uses the ports defined inside the HLSA model and inherits them as shown in figure 6-27.

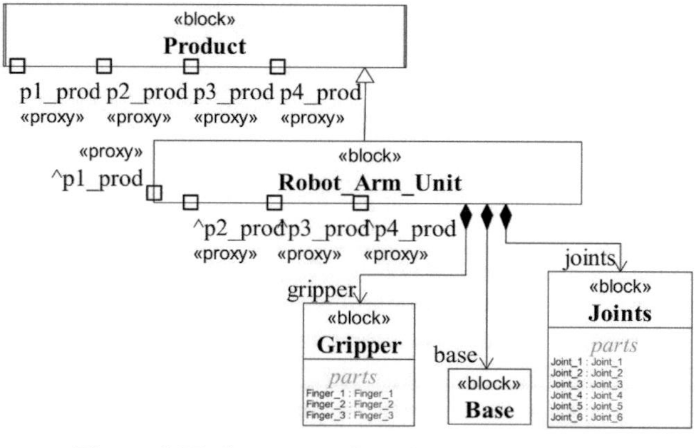

Figure 6-27: Structure of product (robot) model

Along with the inherited ports (indicated with "^"), figure 6-27 also specifies the structure of the robot model i.e. its joints, gripper and base on which the robot is fixed. Similar to the product model, individual models of environment and interaction device are created and saved in their individual project files. Both the interaction device and the environment model inherit the ports from the HLSA model so that the interaction points could be kept consistent throughout the modelling process. The simplified structural definition of the environment model can be seen in figure 6-28 (*Room*) and that of the interaction device can be seen in figure 6-29 (*Interaction_1*). The structural definition starting from figure 6-27 until figure 6-29 are used in describing the behavioural models of product, interaction and environment.

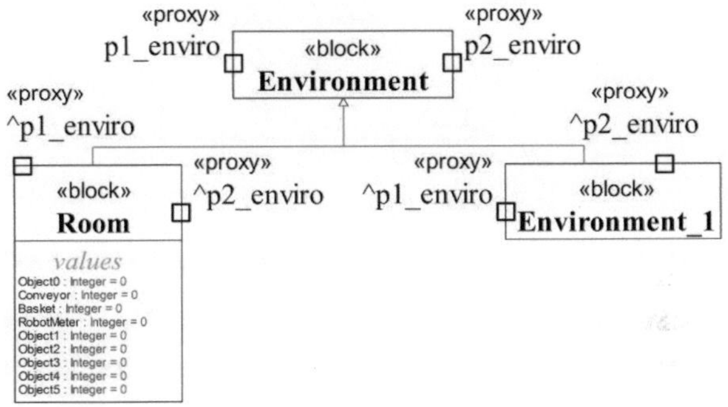

Figure 6-28: Structure of the environment model

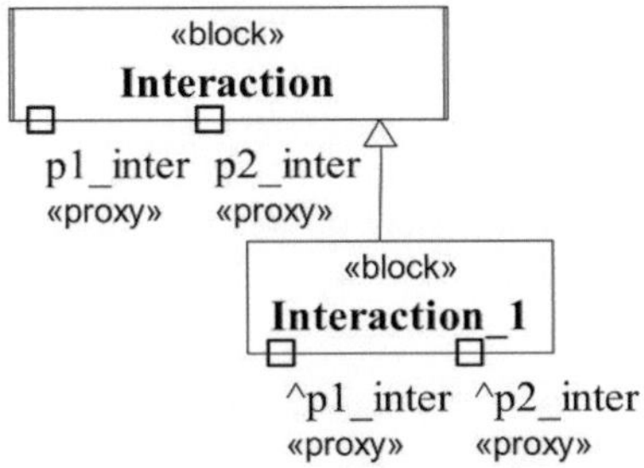

Figure 6-29: Structure of the interaction device model

Figure 6-30 shows the activity diagram in SysML corresponding to *Interaction_1* model and describes its main behaviour in case of manual execution of the robot.

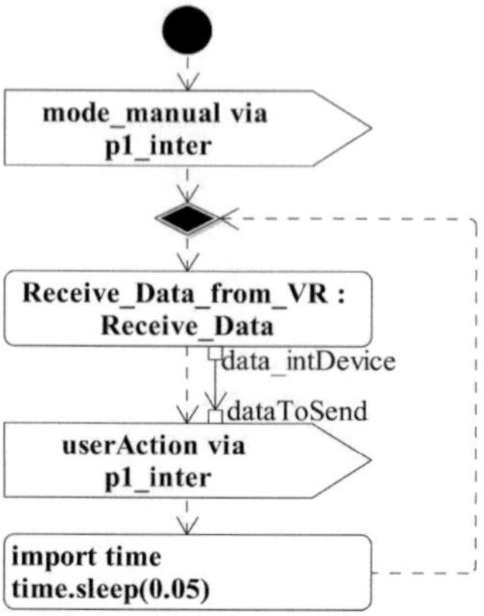

Figure 6-30: Main behaviour of interaction device as ACT

First, a signal "*mode_manual*" is sent to the robot model indicating the mode of operation is manual and later the data coming from the interaction device is constantly evaluated. In the manual mode, the user can manually manipulate the position and orientation of the gripper of the robot by changing the position and orientation of the controller of HMD. The position, orientation and statuses of the button of HMD's controller are analysed, interpreted and extracted data of interest is communicated to the robot model using *userAction* signal over *p1_inter* port.

Figure 6-31 shows the main behaviour of the robot model as STM in SysML.

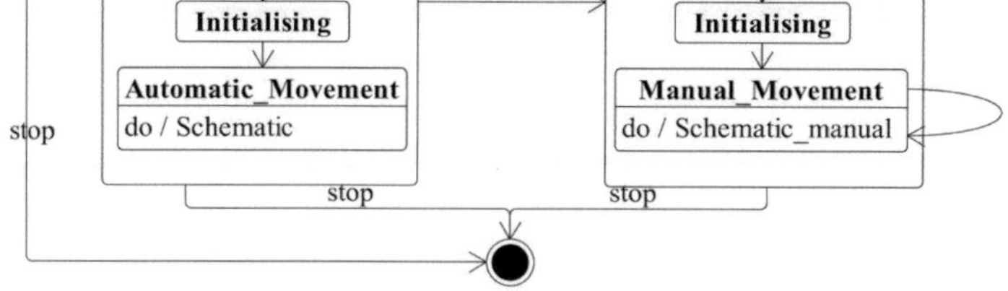

Figure 6-31: Main behaviour of the robot model as STM

After the initialisation, the robot model waits for one of the *mode_auto* or *mode_manual* signals originating from the interaction device model. Two important states of the robot are *Automatic* and *Manual* that corresponds to the mode of its operation. In case of automatic operation, an activity *Schematic* is performed that communicates with the physics model in V-REP and puts the execution in automatic mode. In the automatic model, the conveyer belt feeds the cubes to the default position and the robot grasp the cubes and places them into the basket. To accurately achieve the pick and placement of cubes, inverse kinematics module of V-REP along with the path planning is used.

In the case of the manual operation, an activity *Schematic_manual* is performed endlessly. This activity instructs V-REP to move the position of robotic gripper based on the HMD's controller position and orientation that is received from the interaction device model. The closing of the gripper for grabbing a cube and opening of the gripper to drop the cube into the basket are also handled inside this activity based on the actions performed by the user. The user can directly open and close the gripper using the trigger button (see figure 6-2) of HMD's controller. In this way, the robot can be operated in automatic as well as in manual operation mode that corresponds to an industrial use case in an assembly environment. Furthermore, once the intended robot movement is performed by the V-REP model, a feedback corresponding to the actual position of the gripper, its status and the rotational position of robot joints are received in robot model in SysML. Both the *Schematic* and *Schematic_manual* activities evaluate this feedback and send the updated parameters (e.g. gripper state, position, orientation, joint rotations etc.) to VR-software so that the robot model in VR-scene can be updated.

It can occur during the simulation that the robot hits the conveyor belt or some object in the environment. The V-REP's collision detection module keeps a check on the occurrence of a collision event between two geometric objects and provides continuous feedback to the room model in SysML. Figure 6-32 shows the main behaviour of the environment model as an activity diagram in SysML. This diagram evaluates the feedback about the collisions received from the V-REP model, compares it with the collision status from the last execution run and finally, communicate the changes to VR-software.

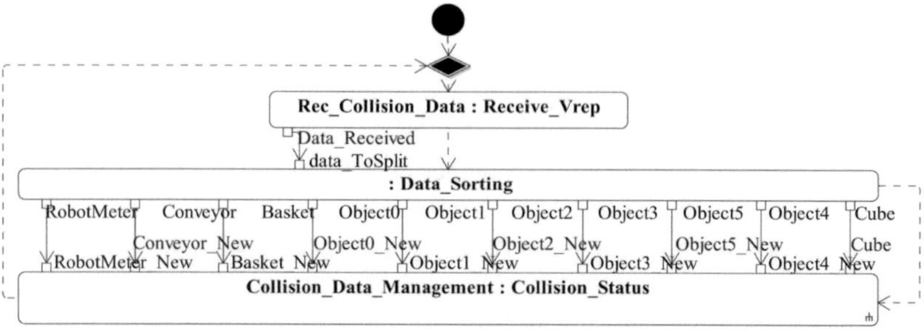

Figure 6-32: Main behaviour of the environment model as ACT

6.2.1.4 Integration of SysML behaviour description with VR

Interaction device integration (interface 1)

To provide interaction device model in SysML with the feedback about the user actions, the controller script (from figure 6-3) is reused that sends out the state of controller buttons, its positions as well its orientation over a UDP connection. In this way, the interface 1 mentioned in figure 5-12 is handled exactly in the same manner as done in 6.1.2.2 by reusing the already developed script.

Updating virtual objects' properties (interface 2)

The updated parameters about the positions of the robot, its gripper, its joints along with the collision statuses are received inside the VR-software by an objects' update script. The same objects' update script (from figure 6-4 & figure 6-5) is reused with minor adjustments. This script receives that updated parameters (position, orientation, rotations, collision etc.) from SysML and updates the objects' properties inside the VR-scene. In this way, the interfaces 2 mentioned in figure 5-12 is implemented by using the same idea as done for the vacuum cleaner example (see 6.1.2.2).

6.2.1.5 Summary

Sub-section 6.2.1 has presented the second VR prototype consisting of 6 degrees of freedom robot working on a conveyor belt that simulates the industrial assembly line scenarios. The behaviour models of the sub-models of product, interaction device and environment are modelled based on the modelling method presented in chapter 4. The modelling of the SysML behaviour models is explained briefly and the behaviour models are integrated with V-REP and VR-software in the same way as explained in chapter 5.

Thus, this prototype

- uses the modelling method developed in chapter 4,
- uses the integration concept as presented in chapter 5 and
- reuses the plugins written in VR-software for the first VR prototype (see 6.1.2.2).

6.2.2 sVR

After presenting the second VR prototype in HMD, the same example scenario is achieved inside sVR using the same SysML behaviour models and V-REP model as used for the case of HMD. The interaction device model in SysML is replaced with the model of Bluetooth controller (see figure 6-14), the already developed positional tracking algorithm (see 6.1.3.2) is reused as it is.

6.2.2.1 Visual model

As the visual model for the second prototype has already been developed inside Unity3d (see 6.2.1.1), it is reused using the same method as described in 6.1.3.1 to create the visual model for sVR. Again Google VR SDK and Google AR Core are used to create the sVR Android application. As a result, a screenshot of the visual model can be seen in figure 6-33.

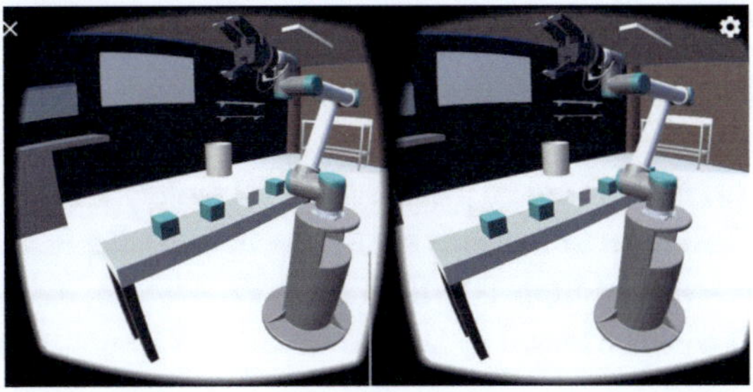

Figure 6-33: VR-scene contents in a smartphone application

6.2.2.2 Integration of SysML behaviour descriptions with VR

The same V-REP model as described in 6.2.1.2 is used in sVR application as well. The SysML behaviour models described in 6.2.1.3 are reused again.

Interaction device integration (interface 1)

The updated data about the Bluetooth controller is sent to SysML behaviour models using the controller script (see figure 6-3) with minor changes. The main change is the use of two physical buttons ("*Two interaction buttons*" in figure 6-14) to control the upwards and downwards movement of the robot gripper. Such modification is necessary to pertain to the fact that the Bluetooth controller does not record and provide the changes in controller height from the floor. In this way, interface 1 mentioned in figure 5-12 is handled exactly in the same manner as done in 6.1.2.2 by reusing the already developed script. Unfortunately, the used controller does not provide any information about its position and orientation. Therefore, the feedback about position and orientation is not included here. This limitation can be removed by using a 6 DoF controller that can provide feedback about its position and orientation as well.

Updating virtual objects' properties (interface 2)

Similar to the case of HMD, the objects' properties are updated by reusing the object update script (from figure 6-4). The TCP/IP connection to receive the updates from SysML behaviour models is changed to UDP connection so that a reliable socket connection can be achieved under Android smartphone. In this way, the interfaces 2 mentioned in figure 5-12 is implemented by using the same idea as done for the vacuum cleaner example (see 6.1.2.2).

6.2.2.3 Summary

Sub-section 6.2.2 has presented the second VR prototype implementation in sVR. The visual model, SysML behaviour models and V-REP developed for achieving VR prototype in HMD are reused to achieve the VR prototype in sVR. Minor modifications are performed to integrate the Bluetooth controller and the connection with SysML is switched from TCP/IP to UDP. On a whole, the VR preparation made for HMD could be reused by large in achieving VR application in sVR.

6.3 Summary

Chapter 6 has presented two VR prototypes that use the method developed in this thesis. The research question 3 mentioned in section 3.5, is answered by reusing the SysML behaviour descriptions, V-REP physics calculation models and geometric models from the VR-software to achieve VR simulation in different VR-systems. These VR simulations have established that SysML behaviour models together with the physics calculations in V-REP build a generic description of complete VR-model that can be reused. Furthermore, these behaviour descriptions are used as a driver of the complete VR simulation in different VR-systems. This proves that SysML behaviour descriptions lie at the core of the simulation process in different VR-systems and thus the sub-question III from section 3.5 is answered.

The next chapter will evaluate the VR applications in CAVE, HMD and sVR for the case of the vacuum cleaner.

7 Evaluation

It is always an extremely challenging task to perform an extensive evaluation of the conducted research and the outcomes that follow. The evaluation of the method for the description of VR-models (see also chapter 4 and 5) that is put forth in this thesis is performed with the help of industrial and academic experts. A holistic evaluation is performed with a focus on the following main categories:

The Utility[87] of the Usability[88]

1. Use of VR in product evaluation
2. Overall satisfaction of experts to VR evaluation in above mentioned VR-systems
3. Concept of context consideration inside the product evaluation

Technical feasibility

4. Feasibility of different VR-systems
    a. CAVE and HMD
    b. HMD and sVR

The focus of the evaluation was put on the developed VR applications (see also 6.1). In this regards, two case studies are performed consisting of a total of 17 test persons including experts from industry as well as academia. Past experience with at least one VR-system was taken as the requirement for the choice of test persons so that the chosen test persons must be familiar with the VR-system. The method used to conduct these case studies as well as the outcomes are described in this chapter.

7.1 Method

The use of VR in the product evaluation, the concept of incorporation of the context of a product inside this evaluation and the feasibility of different VR-systems

---

[87] Utility is justified, when a product can provide the functionality required by the user that is eventually useful for performing the tasks by the user. [Nil93]
[88] Usability is the extent to which a system, a product or a service can be used by certain users in a certain context of use for a certain task so that defined goals of effectivity, efficiency and satisfaction can be achieved [Iso10][Iso98]

for use in product development are tested with the help of empirical evaluations. Next, the design and execution of these evaluations are presented in detail.

7.1.1 Evaluation design

The method presented in this thesis attempts to reduce the overall VR-model preparation effort and thus, attempts to achieve easy incorporation of VR in the current development process. The incorporation of the context (e.g. environment) of a product inside the product evaluation process is also emphasised. Furthermore, the focus is put on the generic behavioural descriptions of a VR-model, that can be reused across different VR-systems. To receive feedback about the usefulness of the presented method, the experts' feedback is obtained using two case studies. A total of 17 experts were involved in both case studies. For ease of understanding, these case studies will be referred to as case study A and case study B. The method used and the execution of both these case studies are performed based on the same mythological concept, however, the times of their execution were different. In each case study, a guided questionnaire is used to collect feedback.

7.1.1.1 Theoretical basis

VR has already found different applications in the industry (see 3.2) and there is even more potential for further applications (see figure 3-9). However, the preparation for VR-models is a difficult task and there is no general method available (see 3.3). Furthermore, in the current product evaluation process, the product remains as the focus of the evaluation and limited consideration of contextual aspects (e.g. life-phase-specific environment of the product) of the evaluation are considered (see table 3 and section 3.3).

The use of any particular VR-system in the industry depends on several factors e.g. the organisation size, turn over, evaluation goal etc. An easy distinction can be made on the base of the size and annual turnover of an industrial organisation. For example, CAVE type VR-system is mainly used by the industrial OEMs whereas HMD finds its application inside small and medium-sized enterprises, game development groups as well as in the domestic market. The current use of VR in the industry is mainly in the design visualisation scenarios and the incorporation of behavioural simulations is limited (see details in 3.2 & 3.3). Smartphone VR (sVR) builds a mobile VR-system, so far its use is limited to the visualisation of 360° videos [Bor17].

Based on the highlighted aspects in this section, the following key areas are identified as the focus of the evaluation:

1. Use of VR in product evaluation
2. Concept of context consideration inside the product evaluation
3. Feasibility of different VR-systems
    a. CAVE and HMD
    b. HMD and sVR
4. Overall satisfaction of experts to VR evaluation in above mentioned VR-systems

These key areas are also kept in consideration while forming the feedback questionnaires and will be discussed in 7.1.2. The evaluation of these aspects is carried out in two case studies and the individual scope of both case studies can be seen in table 11. The greyed out elements are not evaluated inside the mentioned study. Initially, the evaluation focus was only the CAVE and HMD as described in case study A. However, the outcome of case study A suggested an inclination of the test persons towards a mobile and cost-effective VR-system. This motivated the development in sVR. The evaluation of the sVR together with HMD was performed two years later than the first case study and is referred to as the case study B. Therefore, the evaluation process is explained here in exactly the same way as two separate studies with different evaluation scopes.

Table 11: Scope of both case studies and evaluated aspects

Case Study A	Case Study B
1. Use of VR in product evaluation 2. Concept of context consideration in the evaluation 3. Feasibility of different VR-systems a. CAVE and HMD b. HMD and sVR 4. Overall satisfaction of experts to VR evaluation in above mentioned VR-systems	1. Use of VR in product evaluation 2. Concept of context consideration in the evaluation 3. Feasibility of different VR-systems a. CAVE and HMD b. HMD and sVR 4. Overall satisfaction of experts to VR evaluation in above mentioned VR-systems

The highlighted evaluation aspects from table 11 are used in the development of the feedback questionnaire for each case study. For each case study, a separate questionnaire is used that contains the questions addressing the evaluation aspects relevant to the respective study. One evaluation aspect may require multiple questions from the questionnaire to collect the feedback about it. The complete questionnaires corresponding to study A & B can be seen under Annexure D as Questionnaire A and B respectively. The settings for both these case studies are presented next.

7.1.1.2 Setting

Case Study A

As the CAVE type VR-system was involved in the first case study, the experts had to be invited to visit the CAVE VR setup (FASP – see also 2.2.2.1) available at the Technische Universität Ilmenau. Followed by CAVE tests, the tests in HMD (HTC Vive) were also conducted at the Technische Universität Ilmenau in a usability lab on the same day. The important features and hardware setting for the used CAVE are mentioned in 2.2.2.1. HTC Vive with its standard hardware configurations i.e. 2 tracking camera, 2 controllers, HMD itself and a dedication computer for rendering and simulation was used. The usability lab used for conducting the tests provided a 3m X 3m free space for the installation and execution of tests. The CAVE (FASP) and the HTC Vive (Usability Lab) setting used for conducting case study A can be seen in figure 7-1 on the left and middle respectively.

Figure 7-1: CAVE (left), Usability Lab (middle) und Office room (right)

Case Study B

In the case study B, HTC Vive Pro[89] was either used inside an office room environment or was carried directly to the workplace of the test person. The minimum requirement of 3 m X 2.5 m free space was considered while installing HTC Vive Pro on any of the locations. Similar to the HTC Vive Pro setup, sVR consisting of a smartphone[90] with a Bluetooth controller (see also figure 6-14) and VR glasses (Gear VR[91]) for smartphone were used in the same amount of free space at the respective test locations. A sample picture taken during the test in the office room setup can be seen in figure 7-1 on the right.

7.1.1.3 Target group

The target group for the evaluation was the application and research-oriented VR experts. As the goal of the method developed in this thesis is to facilitate the use of VR in product evaluation in industry, the following three groups:

- the product designers/managers/sales personal from industry,
- research assistants from academia and
- the VR-developers

are identified as the target group for conducting these tests. This is because the product designer or manager or salesperson is the eventual user of VR, developers are directly involved in the development of such virtual models and academic researchers can evaluate the application from the scientific perspective.

7.1.1.4 Object

The VR application containing a vacuum cleaner inside a living room environment (explained in chapter 4 and 5) was used as the object. This application as a test object demonstrated a scenario based on real-time vacuum cleaner model simulation inside a virtual living room environment The application allowed control of

[89] HTC Vive Pro is the successor of HTC Vive that offer better resolution, improved cable management, coverage area and few other improvement (see for details https://www.vive.com/de/product/vive-pro/ last accessed 23.03.2020)

[90] Smartphone specifications: 8 core processor (2x 2.6 GHz, 2x 1.92 GHz, 4x 1.8 GHz), 6 GB RAM, 3120x1440 resolution (538ppi), Bluetooth 5.0 and Wifi equipped. Android 10 operating system

[91] Samsung Gear VR: https://www.samsung.com/de/wearables/gear-vr-r323/ [last accessed on 27.03.2020]

the vacuum cleaner model with the help of the VR interaction device. The test persons could move and rotate the vacuum cleaner model as well as could manipulate the angle of its handle using the interaction device. In case of a collision with an object in the living room environment, the change of colour of the colliding object as bright orange was used as a collision indication. The process of achieving this application in CAVE as well HMD is described in detail in chapter 6. The same application was used to achieve one VR application in CAVE type VR-System and one in HMD. The contents of both the applications were the same, however, the differences were present due to different interaction possibilities present in both VR-systems. These differences were unavoidable, as they refer to different technological setups in both VR-systems (see also 2.2.2.1 & 2.2.2.2).

Case study A consisted of applications in CAVE and HMD and was conducted inside the CAVE test environment and the Usability Lab. In figure 7-1, the test environment for the CAVE and the Usability lab can be seen. This interaction device was the flystick[92] device in case of CAVE VR-systems and the game controller (see also figure 6-2) in case of HMD.

Case study B was conducted almost two years after the first case study and hence, based on the received feedback, the application was slightly optimised in terms of speed and product behaviour. This improved application was reused to achieve one VR application in HMD and one in sVR. The example test environment as an office room for this case study can be seen in figure 7-1. The control of the vacuum cleaner was the same as in the former study. The interaction device for HMD was again the game controller and for the sVR, a Bluetooth controller (see also figure 6-14) was used. Besides, the sVR application also possessed the active position tracking algorithm (see 6.1.3.2) that allowed the test persons to move in the virtual living room.

7.1.2 Operationalisation

The basic operationalisation model followed for both case studies can be visualised employing figure 7-2. The first part of the process deals with the introduction to the study and collect the test persons' specific demographical information. The second part contains the variable part of the operationalisation i.e. test object and the tasks performed by the test persons. Although the test object stayed by large

[92] ART Flystick 2: https://ar-tracking.com/products/interaction/flystick-2/ [last accessed on 27.03.2020]

the same in all the studies, the different VR-systems bring in variable aspects in the application evaluation. For instance, the interaction and the control (due to different interaction devices) of the vacuum cleaner model are different in different VR-systems. Similarly, against the given VR-system, the test persons had to perform tasks that can be completed in multiple ways. The third part collects the feedback from the test persons in each case study against individual questionnaires and finally, the feedback is analysed in the fourth part.

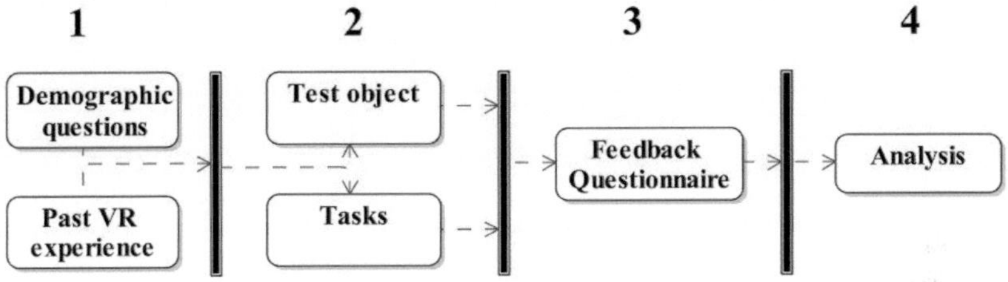

Figure 7-2: Operationalisation model

7.1.2.1 Situation factors

In this operationalisation process model, the experience of the test persons and VR-system related aspects are variables that cannot be influenced directly. The choice of the tasks in each case study is also variable but can be influenced. The variable tasks make it possible to highlight the evaluation related aspects from each case study and therefore, the choice of the tasks is performed keeping in view these aspects (table 11). One situational factor in the case studies was the different interaction device in different VR-systems. For instance, the angle of the vacuum cleaner handle could be changed by varying the height of flystick in CAVE and of the controller in HMD. However, this was not possible in the case of sVR, as the used Bluetooth controller could not measure and provide feedback about its height. Therefore, the angle of vacuum cleaner handle was changed manually by using two buttons (<u>*Two interaction buttons*</u> in figure 6-14) on the Bluetooth controller. Other than this interaction aspects, the chosen tasks and the experience of test persons are considered variables that are discussed next.

Tasks

It is of utmost importance to familiarise the tests persons with the VR applications before collecting the feedback. To ensure that the test persons have not missed any important aspect of the VR application, a total of nine different tasks were used in both studies. These tasks can be seen in table 12 along with the mention of the relevant case study.

Table 12: Tasks performed by test persons in both studies

Tasks	Case Study
1. Please familiarize yourself with the control first	B
2. Please use the vacuum cleaner as a vacuum cleaner in a private living room and try to follow its movement and always stand behind it	A & B
3. Try to clean under the table	A & B
4. Initiate a collision event by making the vacuum cleaner model collide with an environmental object	A & B
5. Change the point of view to visualise the virtual model from different viewpoints	A
6. Try to reach all four corners of the room with the vacuum cleaner (if possible)	A & B
7. Try to look into the virtual model of the vacuum cleaner and identify geometric parts	B
8. Try to clean the space between the table and the sofa	B
9. Try to walk around the table	B

The purpose of the first two tasks was to familiarise the test persons with the control of vacuum cleaner model and with the interaction devices' functionality. Tasks 3 and 4 were targeted to bring the contextual aspect of the evaluation in focus i.e. the evaluation of product functionality along with its context that was in this particular case, product's environment. Task 4 also demonstrated the indication of a collision happening with an environmental object. Task 5 was assigned to familiarise the test persons with different hardware settings and visualisation perspectives available in CAVE and HMD. Once the first five tasks were carried out, the test persons were able to perform task 6 at their own. The purpose of task 6 was to provide test persons with some time so that they may experience the application at their own without the need for any assistance from the organisers.

The task 7 to 9 were added in the case study B because of two reasons. First, both HMD and sVR are very similar in their setup and second, a comparison of the tracking quality of both systems was needed. Therefore, these tasks helped the test persons to estimate a comparison of the tracking quality in both systems. Each of the nine tasks can be practised in multiple ways e.g. the test person can make a collision event with the nearest object or the farthest object. Such a situation makes it difficult to calculate the durations for individual tasks and eventually makes it difficult to perform time-based analysis. Therefore, the focus was not put here on the success rate or time duration of the completion of the tasks, instead, the main goal was to familiarise the test persons with the VR application as well as navigation and interaction possibilities.

7.1.2.2 Person factors

Experience and the current professional position of test persons can on one side greatly influence the test outcome and on the other side, can help to analyse the collected feedback. Therefore, the demographical data about the test persons is collected at the start of each test by using the demographical questionnaires (see under Annexure D). In the choice of test persons, it was made sure that only those test persons are considered who had experience with at least one of VR-system (CAVE, Powerwall, HMD or Smartphone VR) in either private or professional capacity. As a result, a total of 17 test persons (all male gender) participated in both studies including 9 for the case study A and 8 for the case study B. The complete profile of these test persons, their affiliations, age group and past VR experience can be found under Annexure D in table 16. As a whole, there were seven industrial experts affiliated to consumer goods manufacturer, mechanical machinery manufacturer and VR application development. The rest of the ten participants were from academia, among them were researchers from the field of product development, VR developers, programmers etc. The overall experience of the test persons can be seen in figure 7-3 as a graphical depiction.

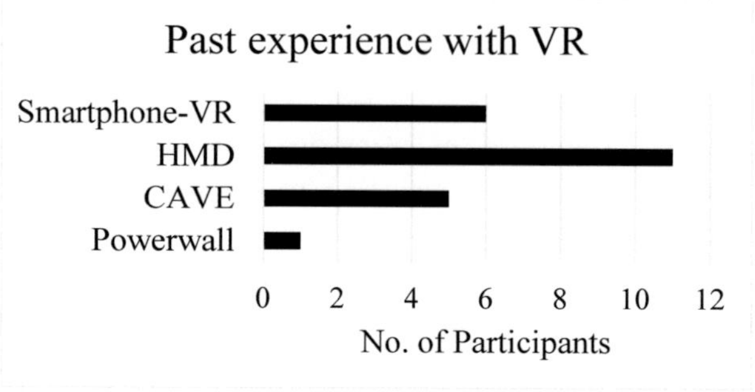

Figure 7-3: Past experience of participants in VR

In an evaluation that is based on the post-test questionnaire, a person related factor is the forgetfulness. The test persons can forget how they felt during the test and fill out the questionnaire on the base of their last feeling. To avoid this, the test persons were given the possibility to go back to test the application again, if they were unsure in answering any question on the questionnaire.

Another crucial person factor occurs in the case study B, as the HMD and sVR are very similar to each other as VR-systems. This leads to the situation that the order in which both applications are shown to the test persons may well influence their final answers. Practically, it is not possible to eliminate this factor, however, change of application order was used as a small workaround so that more objective feedback can be received. This means that the order in which the applications in HMD & sVR were shown to the test persons was flipped for every new test.

7.1.2.3 Scale

The questions in both post-test questionnaires were formulated in a way that a positive or negative answer could be provided. For instance, the test persons could agree with the statement inside the question or can disagree with it. Therefore, a subjective formulation i.e. "strongly agree" to "strongly disagree" was given as answer possibilities for each question. On a four-point Likert scale [Lik32] i.e. strongly agree(++), agree (+), disagree (-) and strongly disagree (--) were the four options against each question. The preference was given to a four-point scale over a five-point scale to avoid middle-value feedbacks and to collect answers that indicate positive or negative tendencies from the test persons.

After laying down the design of the evaluation, the research questions and hypothesis are formulated in the next sub-section 7.1.3.

7.1.3 Research questions and hypothesis

The prime aim of this evaluation was to obtain first feedback from experts affiliated to industry and academia about the developed methodology and the resulting VR applications for use during the product development process. To address these, the following research questions (RQ) were identified to be addressed by the evaluation:

RQ1: Can VR technology be effectively used in industry for product evaluations consisting of not only the geometry but the behaviour of the product as well?

RQ2: Is it important to incorporate the contextual aspects (e.g. environment) of a product during the evaluation?

RQ3: Based on the presented VR applications and application experience in different VR-systems (CAVE, HMD and sVR), is one of the VR-system preferred over the others?

The individual evaluation aspects (table 11) are formulated in the form of hypotheses that shall contribute to answering the above-mentioned research questions. A total of nine hypotheses are formulated and the question numbers (A1-A22 and B1-B17) form feedback questionnaires that address individual hypothesis (H) are also mentioned.

H1: A virtual model is very useful for the evaluation of coherence of product behaviour, its functionality and task practicality (addressed by questions A1 and A2)

H2: It is very important to build the realistic behaviour of the product in VR (addressed by questions A4, A5, A6 and A7)

H3: The consideration of environment model as the product's context is important for product evaluation (addressed by questions A12, A13 and A14)

H4: The collision indication in VR by changing the colour of the collision object is helpful (addressed by question A11)

H5: A virtual model can be detailed to a primitive level for a specific evaluation (addressed by questions A3, A8, A15, B1 and B5)

H6: The vacuum cleaner model possesses realistic behaviour and can be controlled in the desired way (addressed by questions A9, A10, B2, B3, B4, B6 and B7)

H7: HMDs are preferred over CAVE type VR-Systems (addressed by questions A16, A17, A18, A19 and A22)

H8: sVR offers a cost-effective, mobile and better alternative to HMDs (addressed by questions B8, B9, B10, B11, B12, B13, B16 and B17)

H9: The overall evaluation process in VR is very suitable for use in product development (addressed by questions A20 and B14)

The above-mentioned hypotheses are self-explanatory and target the following research fields from this thesis:

1. Product evaluation in VR (see 4 and 5)
    a. Behavioural evaluation (see 4.3.3)
    b. Functional evaluation (see 4.3.3)
    c. Details (see 4.3.3, 5.1, 5.2 and 5.3)
    d. Collision indication (see 5.2 and 5.3)
2. Inclusion of life-phase-specific environment as the context (see 4.1)
3. Comparison of different VR-systems (see 2.2.2, 6 and 7.2)
    a. Immersion
    b. Real (close to reality)
    c. Interaction
    d. Individual preference
4. The usefulness of the overall evaluation method in VR (see 4.1)

After explaining the layout of tests, the rest of this chapter will describe the findings and will also discuss the above-mentioned research fields.

7.1.4 Test execution/organisation

A total of three persons managed the tests conducted under case study A. One person was responsible to handle the technical aspects during simulation, second for assisting the participants in case of questions and the third person noted down any important comments/remarks given by test persons. For the case study B, the tests conducted in the office room setup were handled by a single person, whereas in case of a remote test at test persons' workplace, it was managed by two persons. The execution sequence of both studies can be seen in figure 7-4.

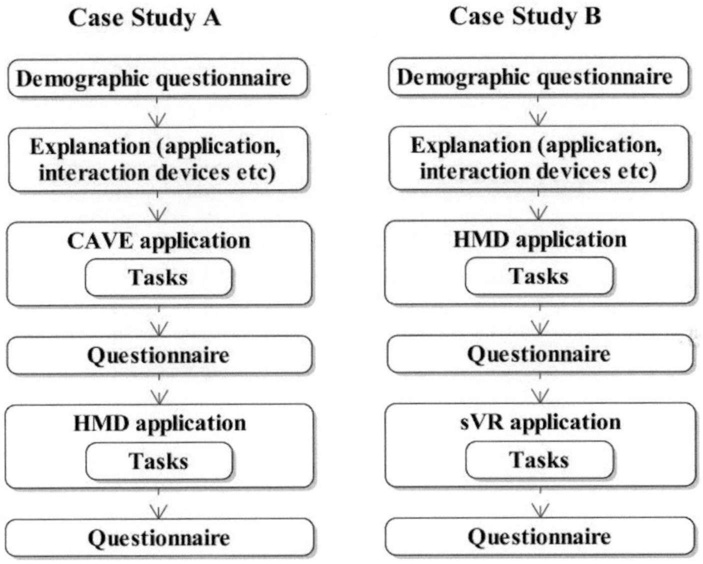

Figure 7-4: Execution sequence

At the start of the tests, the test persons were asked to fill the demographic questionnaire (see in Annexure D) and explanations about basics of the application as well as the use of interaction device were explained. In case study A, the test persons were allowed to experience the application in CAVE VR-system followed by the respective feedback questionnaire. After that, the HMD application was shown followed by the feedback questionnaire. The case study B was carried out in the same way, however, the VR-system here were the HMD and sVR. Furthermore, the sequence of showing each application in HMD and sVR was changed in each test so that more objective feedback can be collected.

7.2 Findings

In this section, the feedback received through the questionnaire will be described and analysed against the hypotheses formulated before conducting the tests.

7.2.1 Pre-processing the data

As a subjective formulation i.e. "strongly agree" to "strongly disagree" on a four-point scale was used, it is necessary to digitalise this formulation so that the received feedback can be demonstrated in the form of easily understandable graphs. Therefore, the four-point scale is digitalised as shown in table 13.

Table 13: Digitalisation of the four-point scale

Subjective formulation	Digital scale
Strongly agree	4
Agree	3
Disagree	2
Strongly disagree	1

All of the feedback data that is presented next in the form of graphs follow the digitalised four-point scale. The vertical axis shows the feedback as mean values (if not mentioned otherwise) of all test persons' answers with standard deviation and a colouring scheme is also used to make the graphics easily understandable. The grey colour scheme is used to plot the bars for general answers about the method developed in this thesis and also for the answers that refer to VR-system independent questions. Blue colour bars are used for CAVE, orange for HMD and green bars are used to plot sVR related answers.

7.2.2 Answers of hypotheses

Hypothesis 1: A virtual model is very useful for the evaluation of coherence of product behaviour, its functionality and task practicality.

This hypothesis is answered by reference to the question A1 and A2 from the feedback questionnaire A that talks about the usefulness of the virtual model for the evaluation of coherence of product functionality and behaviour respectively. The mean values calculated from the nine answers and their standard deviation can be visualised in figure 7-5. The received feedbacks roughly vary from "agree" to "strongly agree" opinion and thus conclude the first hypothesis to be true.

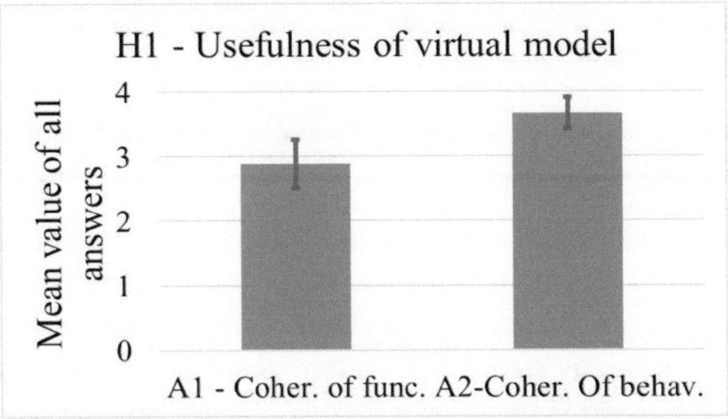

Figure 7-5: Virtual model for the coherence of product functionality and behaviour

Hypothesis 2: It is very important to build the realistic behaviour of the product in VR

The question A4 asks if the vacuum cleaner behaves as expected, A5 asks if the behaviour is realistic, A6 asks if it is realistic to carry out the tasks and A7 asks the importance of building the realistic behaviour in VR. The received feedback (see figure 7-6) indicates that the behaviour of the vacuum cleaner could be improved further and also that realistic behaviour was deemed very important. The relatively low values for A4 to A6 in figure 7-6 refer to an unrealistic/strange turning behaviour of vacuum cleaner model that was observed in the first case study. Based on the feedback, hypothesis 2 can also be concluded true and that it is very important to build realistic behaviour of the product in VR.

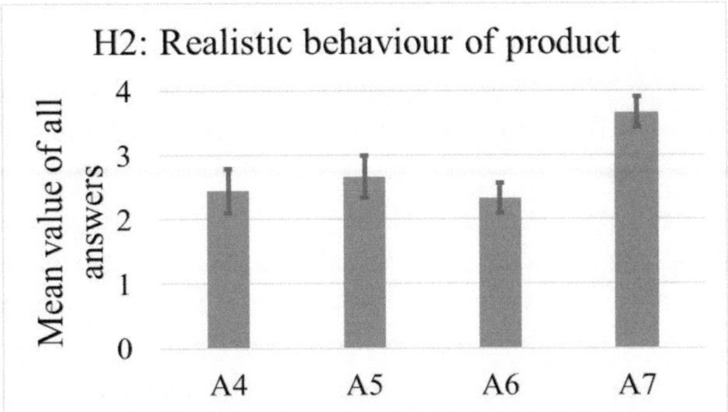

Figure 7-6: Realistic behaviour of VC and the importance of realistic behaviour in general

Hypothesis 3: The consideration of environment model as the product's context is important for product evaluation

Hypothesis 4: The collision indication in VR by changing the colour of the collision object is helpful

Hypothesis 3 can be concluded also as true by evaluating the feedback against question A12, A13 and A14 in figure 7-7. These questions refer to the importance of environment model in evaluating the coherence of product functionality, behaviour and task practicality respectively. Furthermore, hypothesis 4 can also be concluded as true because the collision indication that was implemented as a colour change of colliding object received a higher consent of the test persons (depicted against question A11)

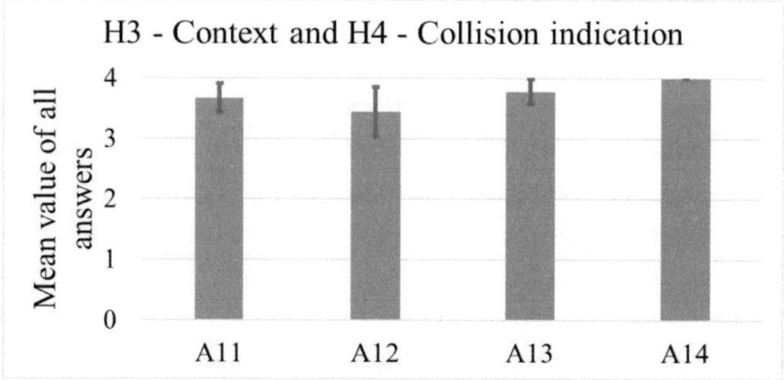

Figure 7-7: Importance of context in evaluation and collision indication

Hypotheses 5: A virtual model can be detailed to a primitive level for a specific evaluation

Questions A3 and A8 ask about the detailed definition of product functions and the level of detail in the product & environment model respectively for the shown application. The lower value against A3 in figure 7-8 refers to the fact that the test persons were not happy about the exclusion of vacuum cleaner power cable in the VR simulation. The inclusion of power cable requires the incorporation of soft body simulation that was left out due to the needed high implementation effort. The questions A15, B1 and B5 refer to the level of detail in the environment and product models as well as the VR-system specific experience. The overall VR-model of product, environment and the control logic was improved after the study A and a slightly optimised version of the application was used in the study B. The received feedback against the questions in figure 7-8 concludes that the hypothesis 5 is false and the virtual model must be as detailed as possible for performing the product evaluation in VR.

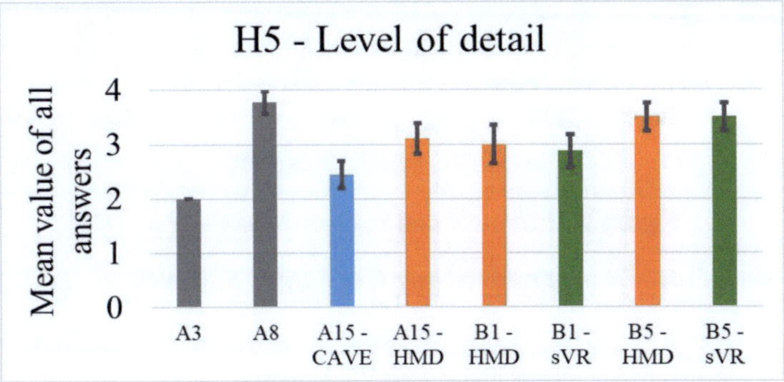

Figure 7-8: Level of details

Hypothesis 6: The vacuum cleaner model possesses realistic behaviour and can be controlled in the desired way

The questions B2, B3 and B6 refer to the realisticness of the vacuum cleaner behaviour in case study B. The question B6 is the replica of A6 and B7 is of A10. Although the trend from figure 7-9 is positive as the answer lies in the upper half, the application can still be improved further. Furthermore, the lower values of B6 and B7 in the case of sVR refer to the low-cost Bluetooth controller that gave an unnatural feeling and the test persons were often irritated by the control. Thus, a better interaction device for sVR can greatly improve the control/interaction of/with the application in sVR. Based on this received data it can be said that hypothesis 6 is not false, however, it is also not completely proved to be true. Against the feedback, this hypothesis can be concluded as partially true.

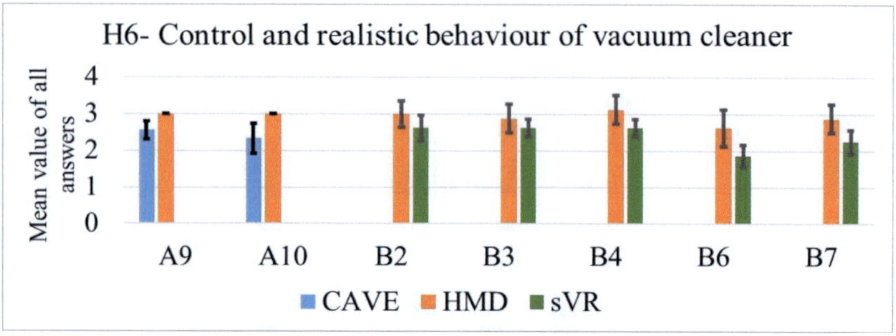

Figure 7-9: Control and realistic behaviour of VC

Hypothesis 7: HMDs are preferred over CAVE type VR-Systems

Question A16 refers to the overall usefulness of the virtual model and A17 to A19 refer to comparative feedback about both VR-systems (CAVE vs HMD) that were tested in study A. Speed, immersion and virtual environment being close to reality in figure 7-10 are rated higher in case of HMD than that for CAVE. Furthermore, figure 7-11 shows the feedback against question A22 that asks the test persons' preference for use of one of the VR-system. Seven persons straight preferred the HMD and two deemed the choice of VR-system to be application depended. Therefore, hypothesis 7 can be concluded true as the HMD is a preference as the VR-system by the test persons.

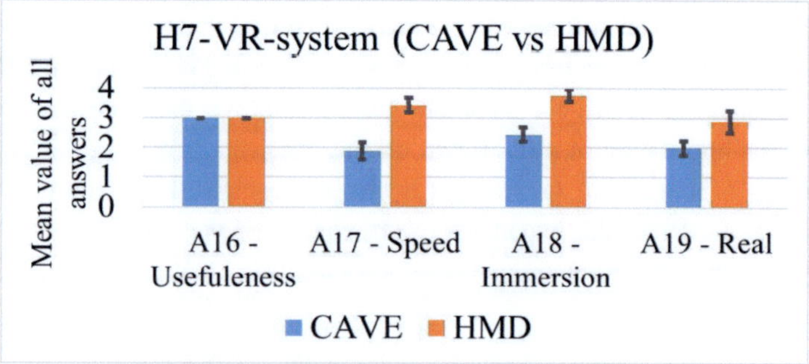

Figure 7-10: VR-system (CAVE vs HMD)

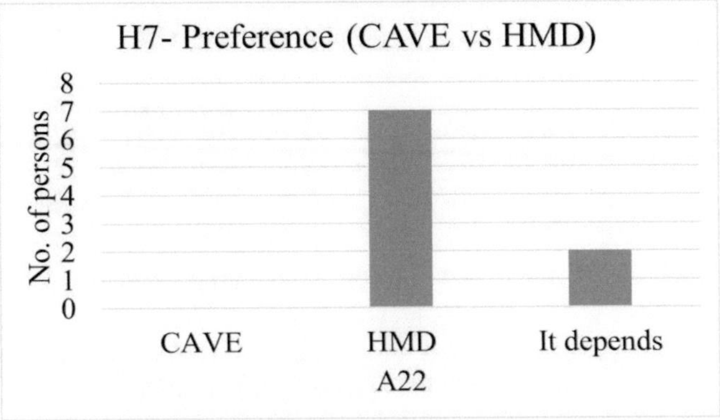

Figure 7-11: VR technology preference

Hypothesis 8: sVR offers a cost-effective, mobile and better alternative to HMDs

The question B8 refers to speed, B9 to B11 to the immersion and being close to reality respectively, B12 to position tracking, B13 to display quality and B16 refers to the installation effort for the VR-system. As depicted graphically in figure 7-12, all of these aspects were rated more positive in case of HMD with the expedition of B16. This suggested that the test persons were more comfortable and inclined towards HMD as the preferred VR-system. However, a lower value of B16 suggests lower installation effort that indicates towards sVR being a more mobile VR-system. Question B17 asked the direct preference of a VR-system (HMD or sVR) and the feedback can be seen in figure 7-13. Multiple answers were possible for this question and four test persons directly preferred HMD over sVR. Five persons considered the choice to be application-specific with one additional person didn't mention any preference. This feedback suggests that hypothesis 8 is false and sVR in the current state is not directly an alternative to HMDs. However, its mobility and current performance make it very suitable for a few VR applications. It was also observed during the tests that the test persons were positively surprised by the capability of sVR. However, the test persons deemed further development as necessary for sVR to be an alternative for HMDs.

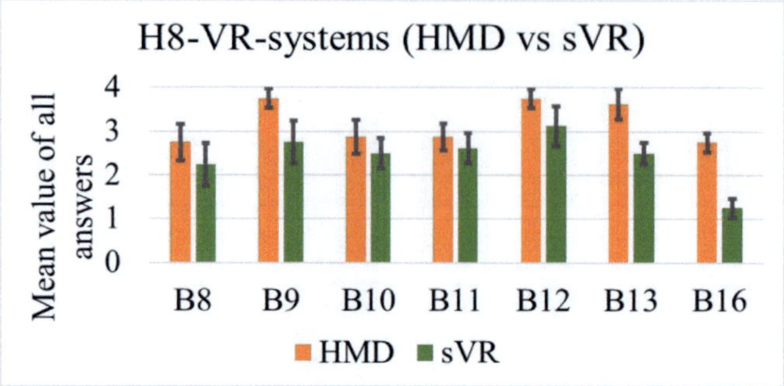

Figure 7-12: VR-system (HMD vs sVR)

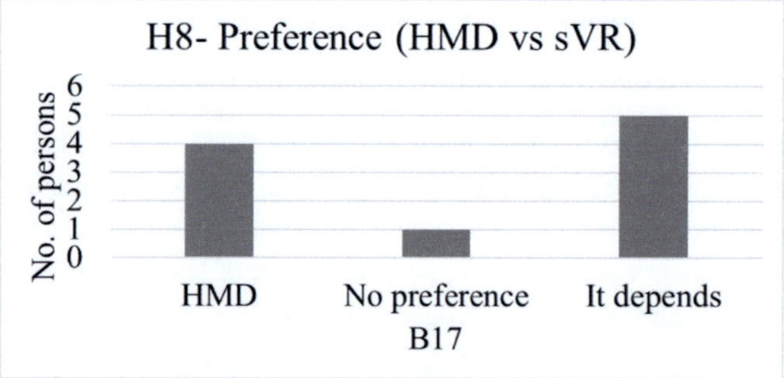

Figure 7-13: VR technology preference

Hypothesis 9: The overall evaluation process in VR is very suitable for use in product development

Finally, the test persons were asked to rate their overall satisfaction from the evaluation process in VR based on the shown applications. The feedback can be seen in figure 7-14 that shows encouraging feedback about the overall evaluation process in VR. Thus, hypothesis 9 can be concluded as true and VR can be identified as an important tool for product evaluation during the product development process.

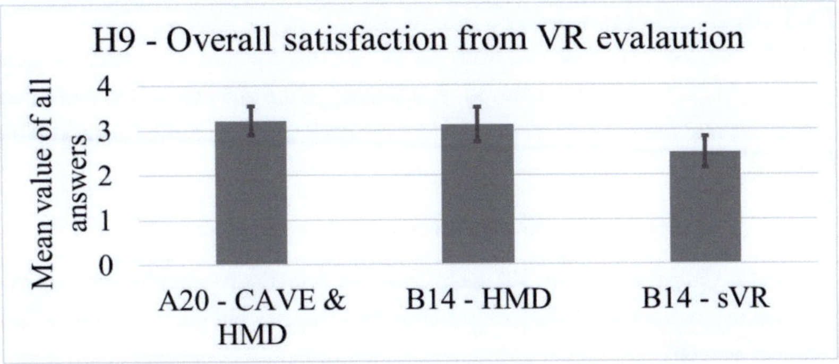

Figure 7-14: Overall satisfaction from VR evaluation

Furthermore, none of the test persons reported any physical or mental disabilities or the effects of motion sickness after conducting the tests under case study A. For case study B, this phenomena can be understood by using the answer of question B15 in figure 7-15.

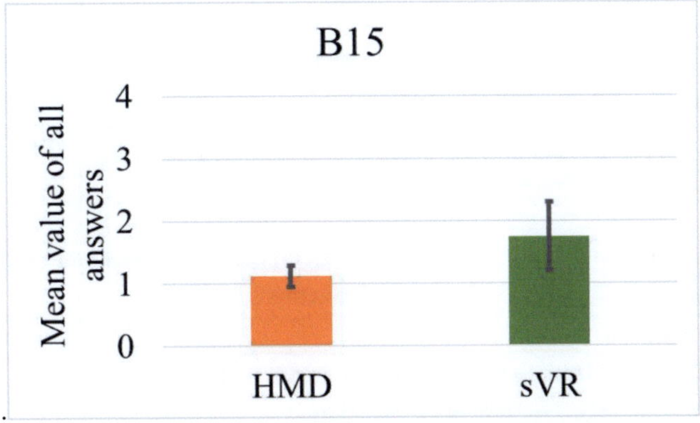

Figure 7-15: Mental or physical limitations after the tests

7.2.3 Answering the research questions

In light of the feedback received from the test persons, the research questions mentioned in 7.1.3 can now be answered. A summary of the outcome of the individual hypothesis and their link to the respective research question can be seen in table 14.

Table 14: Hypotheses involved in answering the RQs

Research Question 1: Can VR technology be effectively used in industry for product evaluations consisting of not only the geometry but the behaviour of the product as well?	
Hypothesis	**Outcome**
H1: A virtual model is very useful for the evaluation of coherence of product behaviour, its functionality and task practicality	True
H2: It is very important to build the realistic behaviour of the product in VR	True
H5: A virtual model can be detailed to a primitive level for a specific evaluation	False
H6: The vacuum cleaner model possesses realistic behaviour and can be controlled in the desired way	Partially (True)
H9: The overall evaluation process in VR is very suitable for use in product development	True
Research Question 2: Is it important to incorporate the contextual aspects (e.g. environment) of a product during the evaluation?	
Hypothesis	**Outcome**
H3: The consideration of environment model as the product's context is important for product evaluation	True
H4: The collision indication in VR by changing the colour of the collision object is helpful	True
Research Question 3: Based on the presented VR applications and application experience in different VR-systems (CAVE, HMD and sVR), is one of the VR-system preferred over the others?	
Hypothesis	**Outcome**
H7: HMDs are preferred over CAVE type VR-Systems	True

H8: sVR offers a cost-effective, mobile and better alternative to HMDs	False

The outcome of hypotheses 1, 2, 5, 6 and 9 can contribute to answering RQ1 and allows to say that VR technology can be used in industry to perform different virtual evaluations of the product. The outcome of hypotheses 3 and 4 suggest that the test persons rate the incorporation of the context of the product (here environment only) as an important value addition to the evaluation process. Thus, RQ2 is also answered. The direct comparison of different VR-systems in hypotheses 7 and 8 has shown that the test persons were inclined towards HMD as the VR-system. Hence, the HMD can be answered as a preferred VR-systems.

7.3 Summary, conclusion and discussion

In this chapter, the results of an empirical evaluation performed on VR application in different VR-systems were presented. Test persons from the industry, as well as academia, were invited to perform the evaluations. The received feedback was encouraging as far as the use of VR for product evaluation is concerned. The overall usefulness of the VR evaluation method developed in this thesis was rated positively. Furthermore, the concept of incorporation of environment model as the product's context in the evaluation was considered as an added value. In terms of VR-system preference, the test persons preferred the HMD over the CAVE in the direct comparison inside the case study A.

The use of Smartphone VR (sVR) as a VR-system in the industry gathered the huge interest of test persons and they were surprised by the capability of sVR for product simulation. However, the test persons rated the interaction possibility to be poor, display quality to be poor and identified a lag comparative to HMD. Due to these aspects, again HMD emerged as a preferred VR-system. However, the possibility of using sVR in product development was not eliminated, the test persons considered it very suitable for remote presentations to the customer, for outdoor application, for quick reviews by the designer during development and in sales for product presentations.

The general feedback from the evaluation and the discussions with the experts showed an inclination towards an HMD based VR-system and the factor related to cost, mobility and minimal need of technical/hardware components were of importance for the test persons. sVR is not mature right now to be a replacement for HMDs but is identified as an important topic for further improvement and research.

8 Conclusion and outlook

In this thesis, a method for the description of VR-models is developed that attempts to reduce the preparation effort by performing a division of the complete VR-model into sub-models (product, actor and environment) and by achieving generic behavioural descriptions of these sub-models. These sub-models can be reused to form different use cases of a product in VR. The thesis contributes to the early evaluation of a product by developing a generic description method based on MBSE approach and by achieving real-time product simulations in VR. The evaluation in VR is not based only on the visualisation, instead, the behaviour of the product is also included. The relevance of the context of a product consisting of the life-phase-specific actor(s) and the environment is also brought to focus and incorporated in the product use cases in VR.

8.1 Summary and discussion

The different VR-systems (CAVE, Powerwall, HMD, Smartphone VR etc.) are already used in industry and their choice for any particular industrial organisation is influenced by factors like organisation size, turnover, developmental goals etc. As a result, different VR-systems find their application in different organisations. The different VR-systems require different VR-software tools and these tools support different programming languages as well as logic. Therefore, a VR-model preparation method based on in VR-software tool programming is not suitable for use during product development (see also 3.3), as it can greatly limit the interoperability and reuse of created VR content in different VR-systems. Therefore, this thesis developed a method based on VR-software independent description method and these descriptions are executed outside of the VR-software.

First, the methodological division of the complete VR-models into sub-models is performed that facilitated the reuse/recombination of these sub-models for the construction of different use cases. Such reuse required generic descriptions of these sub-models and their interactions which was achieved by using MBSE approach with SysML as the modelling language. As a growing interest of industry towards the incorporation of MBSE in the current product development process is present (see also 3.1), the VR-model descriptions based on SysML may pave the way for easy incorporation of VR in the current product development process. The behavioural modelling process in SysML is explained systematically by an example scenario consisting of a vacuum cleaner as a product inside the living room environment. Although the actor model is included inside the methodological foundation of this thesis, the implementation of an actor model is not performed and the VR interaction device is used as its proxy. SysML allowed the description of structural as well as behaviour models, however, the structural models are detailed to the level deemed necessary for the description of the behaviour models. The behaviour models for the sub-models are modelled isolated from each other and direct dependencies are avoided, as direct dependencies in-between sub-models can limit their reuse. The interaction between the sub-models is only allowed over the SysML ports that has facilitated the substitution/reuse of individual sub-models. The behaviour modelling approach in SysML is explained in detail and different execution architectures that are developed and examined throughout the research, are also discussed in detail (see also 4.3.3). As a result, the final modelling approach (see also 4.3.4) is based on the automatic parallel execution of sub-models that has matured over time as a result of continuous research and regular improvements. This modelling approach enabled the automatic parallel behaviour initialisations and generation of different use case configurations of a product for use in VR that were based on the reuse of already created individual descriptions of sub-models. The knowledge gained in developing the final behaviour modelling approach as well as the approach itself is summarised in the form of general-purpose guidelines for developing behavioural descriptions of a dynamic VR-model (see also 4.4).

Although SysML allowed the modelling of the static and dynamic behaviour of the product, it did not offer any possibility to incorporate physical calculations on virtual objects. This is tackled by using a dedicated physics computation software (physics engine) that was also integrated with SysML behaviour models. The integration of the physics engine brought on one side, the much needed physical calculations in the simulation and on the other side, it also contributed to the overall capability of SysML behaviour models. The integration of physics engine and SysML behaviour models is performed in such a way that the physics engine can later also be replaced by any domain-specific simulation tool (see also 5.2). Thus, SysML behaviour models together with a dedicated physics engine built the description of the VR-model that are completely independent of the VR-system used. These descriptions were integrated first with CAVE type VR-system available at the Technische Universität Ilmenau by managing two interface points (see also 5.3 & 5.4). The first interface point received the feedback from the VR-system about the actions performed by the VR-user, interaction device values etc. The second interface point acted as a bridge between SysML behaviour models and VR-software so that the updated properties of virtual objects could be communicated to VR. As a result, a real-time VR simulation of a vacuum cleaner inside a living room was obtained in CAVE type VR-system. Later the same idea was used to achieve similar simulations in an HMD and a Smartphone VR (see also 6.1.2 & 6.1.3). The use of the same behaviour descriptions in different VR-systems validated the generic applicability of the SysML behaviour model and eventually the developed method as well. To further test the presented method, a second VR prototype depicting an industrial robot working on a conveyor belt was developed inside HMD and Smartphone VR (see also 6.2). Again the same SysML behaviour descriptions were used in two different VR-systems with ease.

A partial result of the conducted research is also the use of a Smartphone as a VR device. Smartphone as a VR device lacked direct positional tracking without requiring additional tracking hardware. This was addressed by implementing a positional tracking algorithm that only used the smartphone's camera and achieved direct positional tracking. The achieved positional tracking was tested quantitatively as well as using a survey and encouraging results were obtained, suggesting the diverse application of sVR in the product development process (see 6.1.3.4). One of the goals of the conducted research was to facilitate the incorporation of VR in the current development process, therefore, regular feedbacks were obtained during the research. The final evaluation of vacuum cleaner application inside CAVE, HMD and sVR was performed in the form of empirical evaluations that as well showed encouraging results.

The scope of application of the presented method is neither limited to the size of an industrial organisation nor is it dependent on used VR-systems. The incorporation of HMD and sVR widened the intended audience for the conducted research, as both HMD and sVR offer cost-effective, portable and easily accessible VR solutions. Furthermore, the integration of a physics engine demonstrates the possibility of integrating a domain-specific simulation tool. This means that in the later applications, a domain-specific simulation tool for a particular organisation can be integrated with VR. The domain-specific tool can be a commercial product as well as an in house development, but it must possess an open Application Programming Interface (API) for integration with SysML.

The research questions mentioned in section 3.5 that were the starting point of the conducted research can now be discussed with relevance to the outcomes of this thesis. The first research question was answered by performing a methodological division of the complete VR-model into sub-models of product, actor and environment in section 4.1. The second research question was answered by developing separate isolated behavioural descriptions of the sub-models using SysML in section 4.3. The third research question was addressed by presenting a generic approach for describing VR-models in section 4.2 and achieving generic behavioural descriptions in section 4.3 that can be reused in different VR-systems. The behaviour description methodology was summarized in the form of implementation guidelines (section 4.4) and the generality of the behavioural descriptions was validated by achieving product use case simulations in different VR-systems (chapter 5 and 6). The underlying sub-questions I, II and III were also addressed systematically. The sub-question I was answered by modelling the interaction between the sub-models generically with the help of SysML ports (sub-section 4.3.2). The sub-question II was addressed in chapter 5 by integrating the SysML behaviour descriptions with a dedication physics calculation software. The sub-question III was addressed by achieving VR simulation that used the SysML behavioural description as the core of the simulation process. Furthermore, these behavioural descriptions were used to build two VR prototypes in different VR-systems (chapter 5 and 6) that demonstrated interactive product simulation in VR.

In this way, this thesis has developed a method for the description of VR-models for use in product development to make the early evaluation of product possible in VR. The focus is also put on building the behaviour of the product along with visual representations in VR. The division of complete VR-model in sub-models allow the reuse of these sub-models to build different use cases of a product. Furthermore, VR descriptions developed for a particular VR-system can be reused in different VR-systems. Thus, the VR-model preparation effort is reduced by reusing the already generated VR contents. The VR prototypes validate the method by reusing the VR content developed for one VR-system in other VR-systems as well. Similar to the other scientific researches, the research conducted in this thesis also possesses limitations that are discussed next.

8.2 Limitations and future work

As the presented method is based on a simulation method outside of VR-software, the speed/ data transfer rate with which the data is communicated between VR and SysML behaviour models is very critical. To have a smooth simulation in VR, the updated object parameters must be communicated to VR regularly. Once the updated parameters are received at the VR-software side, they should be immediately implemented. However, the speed of this update depends on the rendering speed of the VR-software and can be critical at times, when a huge data has to be rendered during a VR simulation. It was observed during the simulation that a slower rendering directly affected the simulation speed and as a result, small steps in the movements could be observed. Thus, the speed of communication between SysML and VR-software can be improved further in the future.

Another critical factor is the speed of execution of SysML behaviour models. In the experiments conducted in this thesis, all three sub-models were executed on a single computer that was usually running on 90-95 per cent of its processing capacity. The author believes that the execution speed of the SysML model can be further improved by executing each sub-model on individual computers added to the use of parallel execution architecture. In this way, a parallel computation cluster consisting of multiple computers can be built that may eventually increase the overall application execution speed.

The smartphone used for sVR applications in the evaluation had a pixel density of 538, the maximum resolution of (3120 X 1440), 6 GB of RAM and an 8 core processor. The achieved sVR application possessed 45-60 fps. The smartphones are increasingly equipped with higher hardware specification and even at the time of writing this thesis, there are already some smartphone devices available with even better hardware specifications and performance. Therefore, the display quality and the overall performance of the sVR application can be further improved by using better-equipped smartphones in future. Furthermore, the collaborative VR experience using multiple smartphones can also be investigated.

Unfortunately, it was not possible in this thesis to include a virtual human model as the actor model, as it is a full-fledged research topic in itself. However, the inclusion of a virtual human model can greatly contribute to the usefulness of the overall system, as it can allow the product designer to be the observer of the interplay of actor, product and environment model during product evaluation. Therefore, the inclusion of a virtual human model is identified as the topic of further research.

Literature

[Abi15] ABIDI, Mohamed-Amine; CHEVAILLIER, Pierre; LYONNET, Barbara; KECHICHE, Marwene; BAERT, Patrick and TOSCANO, Rosario: How to Create a New Generation of Industrial Processes Simulation by Coupling Simulation Tools with VR Platforms, in: *28th International Conference on Computer Applications in Industry and Engineering* (CAINE-2015)

[Abi16] ABIDI, Mohamed-Amine; LYONNET, Barbara; CHEVAILLIER, Pierre and TOSCANO, Rosario: Contribution of Virtual Reality for Lines Production's Simulation in a Lean Manufacturing Environment. *International Journal of Computer Theory and Engineering*, 8(3), 182, 2016

[Abu11] ABULRUB, Abdul-Hadi Ghazi; ATTRIDGE, Alex and WILLIAMS, Mark A.: Virtual reality in engineering education, in: *The future of creative learning, IEEE global engineering education conference (EDUCON)* (pp. 751-757). IEEE. (2011)

[Aki19] AKIN, Mahir; GRIMM, Volker; HUSUNG, Stephan; STÖCKELER, Christian; HAHN, Davis and KLEINER, Sven: Beherrschung komplexer mechatronischer Systemvariantens mittels MBSE and SysML, in: *Tag des Systems Engineering (TdSE)* 2019. ISBN: 978-3-9818805-5-7

[Alb14] ALBERS, Albert; LANDES, David; BEHRENDT, Matthias; WEBER, Chrisitan; SIEGEL, Antje and HUSUNG, Stephan: Determination of the Near-Field-Acoustics of Primary Vehicle Sound Sources in Relation to Indoor Pass-by Noise Testing for the Verification of a Virtual Acoustic Vehicle Model, in: *58th Ilmenau Scientific Colloquium*, Ilmenau, 08-12 September 2014

[Ale18] ALEKSANDRAVICIENE, A. and MORKEVICIUS, A.: *"Solution domain", MagicGrid Book of Knowledge- Apractical guide to System Modeling using MagicGrid from No Magic*, pp. 80, Kaunas, Lithuania (2018)

[Alt12] ALT, Oliver: *Modellbasierte Systementwicklung mit SysML*. Carl Hanser Verlag GmbH Co KG, 2012

[Alt20] Atlas Transformation Language,
Url: https://www.eclipse.org/atl/ [last accessed on 30.03.2020]

[Ami16] AMIN, A.; GROMALA, D.; TONG, X.; and SHAW, C.: Immersion in cardboard VR compared to a traditional head-mounted display, in: *International Conference on Virtual, Augmented and Mixed Reality* (pp. 269-276). Springer, Cham (2016)

[And96] ANDREASEN, M.M. and MONTENSEN, N.H.: Basic thinking patterns and working methods for multiple DFX", in: *Proceedings of the 7th Symposium "Design for X"*, Universität Erlangen-Nürnberg, 1996/97, pp. 7-12

[Ang08] ANGERMANN, Frank and RADEMACHER, Martin: Entwicklung und Evaluierung virtueller Prototypen für die Produktentwicklung. *Medienprojekt, Fakultät für Elektrotechnik und Informationstechnik, Fachgebiet Medienproduktion, Technische Universität Ilmenau*, 2008.

[Ang11] ANGERMANN, Anne; BEUSCHEL, Michael; RAU, Martin; WOHLFARTH, Ulrich: MATLAB® – Simulink® – Stateflow® Grundlagen, Toolboxen, Beispiele, in: *7. Auflage. Oldenbourg Verlag München*, ISBN: 978-3-486-70585-0

[Art17] System user manual – ARTtrack, TRACKPACK & DTRack, version 2.13, September 2017. © A. R. T. GmbH, pp. 250-251.

[Bam13] BAMODU, Oluleke; YE, Xuming: Virtual Reality and Virtual Reality System Components, in: *Advanced Materials Research*, vol 765m pp. 1169-1172,2013

[Ban16] BANK, D.; BLUMRICH, F.; KRESS, P. and STÖFERLE, C.: A systems engineering approach for a dynamic co-simulation of a SysML tool and Matlab, in: *2016 Annual IEEE Systems Conference (SysCon)* (pp. 1-6). IEEE, (2016)

[Bat12] BATARSEH, Ola and MCGINNIS, Leon F.: System modeling in sysml and system analysis in arena, in: *Proceedings of the Winter Simulation Conference* (p. 258). Winter Simulation Conference, December, (2012)

[Ber17] BERG, L. P., and VANCE, J. M.: An industry case study: investigating early design decision making in virtual reality. *Journal of Computing and Information Science in Engineering, 17*(1), 011001 (2017)

[Bis92] BISHOP, Gary; BRICKEN, U William and OTHERS: *Research directions in virtual environments*, Report of an NSF Initial workshop. University of North Carolina at Chapel Hill, March 23-24, (1992)

[Bor09] BORDEGONI, M.; CUGINI, U.; CARUSO, G. and POLISTINA, S.: Mixed prototyping for product assessment: a reference framework. *International Journal on Interactive Design and Manufacturing (IJIDeM)*, 3(3), 177-187 (2009)

[Bor17] BORISOV, N.; SMOLIN, A.; STOLYAROV, D.; SHCHERBAKOV, P. and TRUSHIN, V.: The opportunities of applying the 360° video technology to the presentation of cultural events, in: *Interactivity, Game*

Creation, Design, Learning, and Innovation (pp. 256-263). Springer, Cham (2017)

[Bos13] BOSTANCI, E.; KANWAL, N.; EHSAN, S. and CLARK, A. F. "User tracking methods for augmented reality". *International Journel of Computer Theory and Engineering, Vol 5, No 1*, February (2013)

[Bow06] BOWMAN, Doug A.; CHEN, Jian; WINGRAVE, Chadwick A. and OTHERS: New directions in 3D User Interfaces. *The International Journal of Virtual Reality (IJVR)*, vol2(24), pp3-14, (2006)

[Bow08] BOWMAN, D. A.; COQUILLART, S.; FROEHLICH, B.; HIROSE, M.; KITAMURA, Y.; KIYOKAWA, K. and STUERZLINGER, W.: 3d user interfaces: New directions and perspectives, in: *IEEE computer graphics and applications*, *28*(6), pp 20-36, (2008).

[Bra04] BRANDENBURG, Karlheinz; BRIX, Sandra and SPORER, Thomas: Wave field synthesis: New possibilities for large-scale immersive sound reinforcement, in: *Fraunhofer IDMT & Ilmenau Tech University, Mo5 E* 1 (2004).

[Bru10] BRUNO, F.; ANGILICA, A.; COSCO, F. and MUZZUPAPPA, M.: Functional behaviour simulation of industrial products in virtual reality, in: *International symposium on Tools and Methods of competitive engineering, TMCE 2010* (Vol. 2, pp. 763-774). Delft University Technology, Faculty of Industrial Design Engineering (2010)

[Bru13] BRUNO, F., ANGILICA, A., COSCO, F., & MUZZUPAPPA, M.: Reliable behaviour simulation of product interface in mixed reality. Engineering with Computers, 29(3), 375-387 (2013)

[Buc04] BUCHE, C.; QUERREC, R.; DE LOOR, P. and CHEVAILLIER, P.: Mascaret: A pedagogical multi-agent system for virtual environments for training. *International Journal of Distance Education Technologies (IJDET)*, 2(4), 41-61 (2004)

[Bul00] BULLINGER, H. J.; RICHTER, M. and SEIDEL, K. A.: Virtual assembly planning, in: *Human Factors and Ergonomics in Manufacturing & Service Industries*, 10(3), 331-341, (2000)

[Bül14] BÜLTERMANN, R. and BRANDSTÄTTER, M.: Austausch von Modellen zwischen SysML und Modelica zur Simulation, in: *Tag des Systems Engineering*. Hanser, München, S, 289-298, (2014)

[Che12] CHEVAILLIER, P.; TRINH, T. H.; BARANGE, M.; DE LOOR, P.; DEVILLERS, F.; SOLER, J. and QUERREC, R.: Semantic modeling of virtual environments using mascaret, in: *5th Workshop on Software*

Engineering and Architectures for Realtime Interactive Systems (SEARIS) (pp. 1-8), IEEE, March, (2012)

[Cho15] CHOI, S.; JUNG, K. and NOH, S. D.: Virtual reality applications in manufacturing industries: Past research, present findings, and future directions, in: *Concurrent Engineering: Research and Application*, 23(1), 40-63, 2015.

[Cru92] CRUZ-NEIRA, C.; SANDIN, D. J.; DEFANTI, T. A.; KENYON, R. V. and HART, J. C.: The CAVE: audio visual experience automatic virtual environment, in: *Communications of the ACM*, 35(6), 64-73 (1992)

[Dän14] DÄNZER, M.; KLEINER, S.; LAMM, J. G.; MOESER, G.; MORANT, F.; MUNKER, F. and WEILKIENS, T.: Funktionale Systemmodellierung nach der FAS-Methode: Auswertung von vier Industrieprojekten, in: *Tag des Systems Engineering* 2014, 75-84 (2014)

[Des01] DESEL, J. and JUHÁS, G.: “What Is a Petri Net?” Informal Answers for the Informed Reader, in: *Unifying Petri Nets* (pp. 1-25). Springer, Berlin, Heidelberg, (2001).

[Eig15] EIGNER, M.; DICKKOPF, T.; SCHULTE, T. and SCHNEIDER, M.: *„mecPro^2 – Entwurf einer Beschreibungssystematik zur Entwicklung cybertronischer Systeme mit SysML“*, in: Schulze, S.; Muggeo, C. (Hrsg.): Tag des Systems Engineering, Hanser Verlag, München 2015, S. 163-172. – ISBN: 978-3-446-44729-5

[Eig16] EIGNER, Martin; DICKOPF, Thomas; APOSTOLOV, Hristo; SCHAEFER, Patrick; FAIßT, Karl-Gerhard et al.: System Lifecycle Management: Initial Approach for a Sustainable Product Development Process Based on Methods of Model Based Systems Engineering, in: *Shuichi Fukuda; Alain Bernard; Balan Gurumoorthy; Abdelaziz Bouras. 11th IFIP International Conference on Product Lifecycle Management (PLM), Jul 2014, Yokohama, Japan. Springer, IFIP Advances in Information and Communication Technology, AICT-442, pp.287-300, 2014, Product Lifecycle Management for a Global Market* (2014) <10.1007/978-3-662-45937-9_29>

[Eig17a] EIGNER, M., DICKOPF, T., SCHNEIDER, M. and SCHULTE, T.: mecPro¸-A holistic concept for the model-based development of cybertronic systems, in: *DS 87-3 Proceedings of the 21st International Conference on Engineering Design (ICED 17) Vol 3: Product, Services and Systems Design*, Vancouver, Canada, 21-25.08. 2017 (pp. 379-388) (2017)

[Eig17b] EIGNER, M., KOCH, W., & MUGGEO, C.: Modellbasierter Entwicklungsprozess cybertronischer Systeme, in, *Der PLM-unterstützte Referenzentwicklungsprozess für Produkte und Produktionssysteme*. Springer Vieweg (2017)

[Elc11] EL-CHAAR, J., BOER, C. R., PEDRAZZOLI, P., MAZZOLA, S. and DAL MASO, G.: Interactive 3D virtual environments for industrial operation training and maintenance, in: *The Proceedings of 2011 9th International Conference on Reliability, Maintainability and Safety* (pp. 1376-1381). IEEE (2011)

[Epp18] EPPLE, Jochen and FROTSCHER, Ralf: Introduction of Model-based Methods into an Established Large-scale Systems Engineering Process, in: *EMEASEC 2018 / TdSE 2018*, Berlin, Germany (2018)

[Est07] ESTEFAN, Jeff A.: Survey of model-based systems engineering (MBSE) methodologies. International Council on Systems Engineering (INCOSE). INCOSE-TD-2007-003-02. Rev. B. Seattle, WA, USA. [last accessed 30.03.2020] http://www.omgsysml.org/MBSE_Methodology_Survey_RevB.pdf

[Exn19] EXNER, Konrad: Prototyping von Produkt-Service Systemen und Smart Services in der Konzeptphase des Entwicklungsprozesses. Fraunhofer Verlag (2019). ISBN: 978-3-8396-1457-0

[Fan17] FANG, W., ZHENG, L., DENG, H. and ZHANG, H.: Real-Time Motion Tracking for Mobile Augmented/Virtual Reality Using Adaptive Visual-Inertial Fusion, in: *Sensors* (Basel, Switzerland), 17(5), 1037, 2017. doi:10.3390/s17051037 (2017)

[FAS20] OOSE eG; :EM AG ; IPEK - Institut für Produktentwicklung am KIT ; Helmut-Schmidt-Universität: FAS4M – Functional Architectures of Systems for Mechanical Engineers.URL http://fas4m.de. [last accessed on 30.03.2020]

[Fol10] FOLLMER, M.; HEHENBERGER, P.; PUNZ, S. and ZEMAN, K.: Using SysML in the product development process of mechatronic systems, in: *DS 60: Proceedings of DESIGN 2010, the 11th International Design Conference*, Dubrovnik, Croatia (pp. 1513-1522) (2010)

[Fou12] FOURES, D.; ALBERT, V.; PASCAL, J. C. and NKETSA, A.: Automation of SysML activity diagram simulation with model-driven engineering approach, in: *Proceedings of the 2012 Symposium on Theory of Modeling and Simulation-DEVS Integrative M&S Symposium* (p. 11). Society for Computer Simulation International, March, (2012)

[Fri98] FRITZSON, Peter and ENGELSON, Vadim: Modelica—A unified object-oriented language for system modeling and simulation, in: *European Conference on Object-Oriented Programming*. Springer, Berlin, Heidelberg, (1998)

[Gau15] GAUSEMEIER, J., DUMITRESCU, R., STEFFEN, D., CZAJA, A., WIEDERKEHR, O. and TSCHIRNER, C.: Systems Engineering in industrial practice. Heinz Nixdorf Institute, University of Paderborn, Faculty of Product Engineering Fraunhofer Institute for Production Technology IPT –Project Group Mechatronic Systems Design UNITY AG. Wentker druck GmbH, Paderborn, (2015)

[Gor17] GÓRSKI, F., BUŃ, P., WICHNIAREK, R., ZAWADZKI, P. and HAMROL, A.: Effective Design of Educational Virtual Reality Applications for Medicine using Knowledge-Engineering Techniques. *Eurasia Journal of Mathematics, Science & Technology Education*, 13(2), 2017.

[Gro15] GROSS, J. and MUKHERJEE, R.: Integrating Multibody Simulations with SysML, in: *ASME 2015 International Design Engineering Technical Conferences and Computers and Information in Engineering Conference* (pp. V006T10A038-V006T10A038). American Society of Mechanical Engineers, August, (2015)

[Har15] HART, Laura E.: Introduction to model-based system engineering (MBSE) and SysML, in: *Delaware Valley INCOSE Chapter Meeting*, Ramblewood Country Club, Mount Laurel, New Jersey (2015)

[Hol19] HOLOPAINEN, J., MATTILA, O., PARVIAINEN, P., PÖYRY, E. and TUUNANEN, T.: Enabling Sociability When Using Virtual Reality Applications: A Design Science Research Approach, in: *Proceedings of the Annual Hawaii International Conference on System Sciences*. University of Hawai'i at Manoa (2019)

[Htc20] Vive PRE User Guide, online: https://www.htc.com/managed-assets/shared/desktop/vive/Vive_PRE_User_Guide.pdf [last accessed 30.03.2020]

[Hum12] HUMMEL, Johannes; WOLFF, Robin; STEIN, Tobias; GERNDT, Andreas and KHULEN, Torsten: An evaluation of open sources physics engines for use in virtual reality assembly simulations, in: *International Symposium on Visual Computing* (pp. 346-357). Springer, Berlin, Heidelberg, (2012)

[Hus09] HUSUNG, S., WEBER, C. and GRAMSTAT, S.: Simulation of acoustical product properties for technical systems in virtual environments, in: *DS 58-5: Proceedings of ICED 09, the 17th International Conference on Engineering Design, Vol. 5, Design Methods and Tools (pt. 1),* Palo Alto, CA, USA, 24.-27.08. 2009 (pp. 85-96) (2009)

[Hus11] HUSUNG, Stephan: Simulation akustischer Produkteigenschaften unter Nutzung der Virtual Reality während der Produktentwicklung. Universitätsbibliothek Ilmenau, (2011)

[Hus14] HUSUNG, Stephan; SIEGEL, Antje; WEBER, Christian: Acoustical investigations in Virtual Environments for a car passing application, in: *Proceedings of the ASME 2014 International Design Engineering Technical Conferences & Computers and Information in Engineering Conference IDETC/CIE 2014*, August 17-20, 2014, Buffalo, New York, USA (2014)

[Hus15] HUSUNG, Stephan; SIEGEL, Antje; WEBER, Christian; MÜLLER, Bernd; ALBERS, Albert; LANDES, David and BEHRENDT; Matthias: Audio-visuelle Simulation der Fahrzeugvorbeifahrt unter Nutzung von VR und Prüfstandsmessungen, in: *12. Paderborner Workshop Augmented & Virtual Reality in der Produktentstehung*, (2015)

[Iee17] IEEE Standard VHDL Analog and Mixed-Signal Extensions- IEEE 1076.1-2017

[Inc07] International Council On Systems Engineering (INCOSE): Systems Engineering Vision 2020, INCOSE_TP-2004-004-02, Version/Revision: 2.03, September 2007

[Inc15] Systems Engineering Handbook: A Guide for System Life Cycle Processes and Activities. Fourth edition. INCOSE-TP-2003-002-04, 2015.

[Inc20] Definition of Systems Engineering: url: https://www.incose.org/systems-engineering, (last accessed 30.03.2020)

[Iso10] DIN EN ISO 9241-210:2010. Ergonomics of human-system interaction. Part: 210 - Human-centred design for interactive systems, 9241-210 (2010)

[Iso98] DIN EN ISO 9241-11:1998. Ergonomische Anforderungen für Bürotätigkeiten mit Bildschirmgeräten - Teil 11: Anforderungen an die Gebrauchstaug-lichkeit; Leitsätze, Berlin: Beuth Verlag, 1998

[Iso17] ISO/IEC 19514:2017 - Information technology -- Object management group systems modeling language (OMG SysML), https://www.iso.org/standard/65231.html [last accessed on 30.03.2020]

[Jay99] JAYARAM, S., JAYARAM, U., WANG, Y., TIRUMALI, H., LYONS, K. and HART, P: VADE: a virtual assembly design environment, in: *IEEE computer graphics and applications*, 19(6), 44-50, (1999)

[Kir16] KIRSCH, Lucas; MÜLLER, P.; EIGNER, M.; MUGGEO, C.: SysML-Modellverwaltung im PDM/PLM-Umfeld, in: *Tag des Systems Engineering:2016*, Hanser, München 333-342. ISBN: 9783446451261 (2016)

[Kle13] KLEINER, Sven; KRAMER, Christoph: Model based design with systems engineering based on RFLP using V6, in: *Smart Product Engineering*. Springer, Berlin, Heidelberg, 2013. S. 93-102 (2013)

[Kle16] KLEINER, S. and HUSUNG, S.: Model Based Systems Engineering: Prinzipien, Anwendung, Beispiele, Erfahrung und Nutzen aus Praxissicht, in: *Tag des Systems Engineering*, pp. 13-22, (2016)

[Koe04] Koenig, N., & Howard, A.: Design and use paradigms for gazebo, an open-source multi-robot simulator, in: *2004 IEEE/RSJ International Conference on Intelligent Robots and Systems* (IROS)(IEEE Cat. No. 04CH37566) (Vol. 3, pp. 2149-2154). IEEE, (2004)

[Köp09] KÖPPEN, V., SIEGMUND, N., SOFFNER, M. and SAAKE, G.: An architecture for interoperability of embedded systems and virtual reality, in: *IETE Technical Review*, 26(5), 350-356 (2009)

[Lav17] LAVIOLA JR, J. J., KRUIJFF, E., MCMAHAN, R. P., BOWMAN, D. and POUPYREV, I. P.: 3D user interfaces: theory and practice, in: *Second edition. Addison-Wesley Professional 2017.* ISBN: 978-0-13-403442-4 (2017)

[Lew90] LEWIS, C., POLSON, P. G., WHARTON, C., & RIEMAN, J.: Testing a walkthrough methodology for theory-based design of walk-up-and-use interfaces, in *Proceedings of the SIGCHI conference on Human factors in computing systems* (pp. 235-242). ACM (1990).

[Lie17] LIEBAL, Andreas; MAHBOOB, Atif; WEBER, Christian; KRÖMKER, Heidi: CPM/PDD-basierter Ansatz für Produktevaluation in Virtual Reality (VR), in: *13. Mageburger Machinenbau – Tage (MMT2017)*, Magdeburg (2017)

[Lik32] LIKERT, R. (1932). A technique for the measurement of attitudes. Archives of Psychology, 22(140), 1–55.

[Lor16] LORENZ, M., SPRANGER, M., RIEDEL, T., PÜRZEL, F., WITTSTOCK, V., & KLIMANT, P.: CAD to VR–a methodology for the automated conversion of kinematic CAD models to virtual reality, in: *Procedia CIRP*, *41*, 358-363, (2016)

[Lus97] LUCKAS, V., & BROLL, T.: CASUS; an object-oriented three-dimensional animation system for event-oriented simulators, in: *Proceedings. Computer Animation'97 (Cat. No. 97TB100120)* (pp. 144-150). IEEE (1997)

[Mac15] MACISAAC, D.: Google Cardboard: A virtual reality headset for $10?, in: *The Physics Teacher*, *53*(2), 125-125. 2015. URL: https://doi.org/10.1119/1.4905824 [last accessed on 09.03.2020].

[Mah16] MAHBOOB, Atif; LIEBAL, Andreas; HUSUNG, Stephan; WEBER, Christian and KRÖMKER, Heidi: Konzept für ein kontextabhängiges VR-Modell und die Verhaltensbeschreibung in SysML am Produktbeispiel, in: *Jahrestagung der GI VR/AR 2016*, Bielefeld, ISBN: 978-3-8440-4718-9 (2016)

[Mah17a] MAHBOOB, Atif; WEBER, Christian; HUSUNG, Stephan; LIEBAL, Andreas and KRÖMKER, Heidi: Model Based Systems Engineering (MBSE) approach for configurable product use-case scenarios in Virtual Environments, in: *21st International Conference on Engineering Design (ICED17)*, 21-25.08.2017, Vancouver (2017)

[Mah17b] MAHBOOB, Atif; LIEBAL, Andreas; HUSUNG, Stephan; WEBER, Christian and KRÖMKER, Heidi: A method for fast and easy configuration of Virtual Reality (VR) Environments for Analysis of Technical Systems, in: *59th – Ilmenau Scientific Colloquium (IWK)*, Ilmenau (2017)

[Mah18a] MAHBOOB, Atif; HUSUNG, Stephan; WEBER, Christian; LIEBAL, Andreas and KRÖMKER, Heidi: SYSML behaviour models for description of Virtual Reality environments for early evaluation of a product, in: *International Design Conference – Design 2018*, Faculty of Mechanical Engineering and Naval Architecture, University of Zagreb, (2018)

[Mah18b] MAHBOOB, Atif; HUSUNG, Stephan; WEBER, Christian; LIEBAL, Andreas and KRÖMKER, Heidi: An approach for building product use-case scenarios in different virtual reality environments, in: *Proceedings of ASME 2018 International Design Engineering Technical Conferences & Computers and Information in Engineering Conference (IDETC/CIE 2018)*, Quebec City, Quebec, Canada (2018)

[Mah19a] MAHBOOB, Atif; HUSUNG, Stephan; WEBER, Christian; LIEBAL, Andreas and KRÖMKER, Heidi: Modellbasierter Systems Engineering Ansatz zur effizienten Aufbereitung von VR-Szenen, in: *EEE2019 – Konferenz Entwerfen Entwickeln Erleben in Produktentwicklung und Design 2019 Band 2*, Dresden, (2019)

[Mah19b] MAHBOOB, A., HUSUNG, S., WEBER, C., LIEBAL, A. and KRÖMKER, H.: The Reuse of SysML Behaviour Models for Creating Product Use Cases in Virtual Reality, in: *Proceedings of the Design Society: International Conference on Engineering Design*, 1(1), 2021-2030. doi:10.1017/dsi.2019.208 (2019)

[Mah19c] MAHBOOB, Atif; HUSUNG, Stephan; WEBER, Christian; LIEBAL, Andreas and KRÖMKER, Heidi: Smartphone as a Stand-Alone Device for Rendering, Visualization and Tracking for Use During Product Development in Virtual Reality (VR), in: *International Design Engineering Technical Conferences and Computers and Information in Engineering Conference* (Vol. 59179, p. V001T02A079), *ASME* (2019)

[Mch08] MCHENRY, Kenton and BAJCSY, Peter: An overview of 3d data content, file formats and viewers, in: *National Center for Supercomputing Applications*, 1205, (2008)

[Mel01] WOOD, Robert B. and HOWELLS, Pete J.: MELZER, James E and MOFFIT, K: Head-mounted display, in: *The Avionics Handbook*. (2001)

[Men17] MENDEZ, Roberto Lopez: Mobile inside-out VR Tracking, now readily available on your phone with Unity. [online] https://blogs.unity3d.com/2017/10/18/mobile-inside-out-vr-tracking-now-readily-available-on-your-phone-with-unity/ [last accessed 30.03.2020]

[Min95] MINE, Mark R. Virtual environment interaction techniques, in: *UNC Chapel Hill CS Dept*, (1995)

[Mod17] Modelica – A Unifiied Object-Oriented Language for Systems Modeling, Language Specification V 3.4. Modelica Association. April 2017. https://www.modelica.org/documents/ModelicaSpec34.pdf [last accessed on 30.03.2020]

[Moe15a] MOESER, G., Example on „Usage of Free Sketches in MBSE ", IPEK-Insitute of Product Engineering at Karlsruhe Insitute of Technology (KIT), Karlsruhe, Germany, 2015.

[Moe15b] MOESER, G.; KRAMER, C.; GRUNDEL, M.; NEUBERT, M.; KÜMPEL, S.; SCHEITHAUER, A.; KLEINER, S.; ALBERS, A.; "Fortschrittsbericht zurmodellbasierten Unterstützung der Kontrukteurstätigkeit durch FAS4M", in: *Tag des Systems Engineering, Ulm*, November 11-13, 2015, Carl Hanser Verlag, München, pp. 69–78. https://doi.org/10.3139/9783446447288.008) (2015)

[Moe16] MOESER G., Kurzversion FAS4M Methodenguideline, 2016. URL: http://nbn-resolving.org/urn:nbn:de:swb:90-562083 [last accessed on 30.03.2020].

[Mue02] MUELLER-WITTIG, W., JEGATHESE, R., SONG, M., QUICK, J., WANG, H., & ZHONG, Y. (2002, December). Virtual factory: highly interactive visualisation for manufacturing, in: *Proceedings of the 34th conference on Winter simulation: exploring new frontiers* (pp. 1061-1064). Winter Simulation Conference (2002)

[Nam19] NAM, S., KO, G., SUH, K. W. and KWON, J.: User Experience-and Design-Oriented Virtual Product Prototyping System, in: *11th International Conference on Knowledge and Smart Technology (KST)* (pp. 116-120), IEEE (2019)

[Nil93] NIELSON, Jakob: *Usability Engineering*. Morgan Kaufmann, San Francisco (1993). ISBN: 1-12-518406-9

[Nou14] Nogueira, L.: Comparative analysis between gazebo and v-rep robotic simulators, in: *Seminario Interno de Cognicao Artificial-SICA*, 2014, 5.

[Nov11] NOVAK-MARCINCIN, J., BRAZDA, P., JANAK, M., & KOCISKO, M.: Application of virtual reality technology in simulation of automated workplaces, in: *Tehnički vjesnik*, *18*(4), 577-580 (2011)

[Omg17a] OMG-SysML: OMG Systems Modeling Language (OMG SysML™, Version 1.5): https://www.omg.org/spec/SysML/1.5/PDF [last accessed on 30.03.2020].

[Omg17b] OMG-UML: OMG Unified Modeling Language (Version 2.5.1). Object Management Group. 2017 [https://www.omg.org/spec/UML/ last accessed 30.03.2020]

[Omg18a] OMG-fUML: Semantics of a Foundational Subset for Executable UML Models (fUML). Version1.4. Object Management Group Document Number: formal/2018-12-01. https://www.omg.org/spec/FUML/1.4/ [last accessed 30.03.2020]

[Omg18b] OMG-SysML: OMG Systems Modeling Language (Version 1.6). Object Management Group. October 2018. https://www.omg.org/spec/SysML/1.6/Beta1/PDF [last accessed 30.03.2020]

[Pap17] PAPACHRISTOS, N. M.; VRELLIS, I. and MIKROPOULOS, T. A.: A comparison between oculus rift and a low-cost smartphone VR headset: immersive user experience and learning, in: *IEEE 17th International Conference on Advanced Learning Technologies (ICALT)* (pp. 477-481), IEEE (2017)

[Par08] PARK, H., SON, J. S. and LEE, K. H.: Design evaluation of digital consumer products using virtual reality-based functional behaviour simulation. *Journal of Engineering Design*, 19(4), 359-375, (2008)

[Pea12] PEARCE, Paul and MATTHEW, Hause: ISO-15288, oosem and model-based submarine design. (2012)

[Pra18] PRATAMA, L. A., & DOSSICK, C. S.: Workflow in Virtual Reality Tool Development for AEC Industry, in: *Advances in Informatics and Computing in Civil and Construction Engineering* (pp. 297-306). Springer, Cham (2018)

[Rad14] RADEMACHER, M. H.: Virtual Reality in der Produktentwicklung: Instrumentarium zur Bewertung der Einsatzmöglichkeiten am Beispiel der Automobilindustrie. Springer-Verlag, Page16 (2014)

[Ric17] RICKS, Michael; ENGELMANN, Michael; KLEINER, Sven; HUSUNG, Stephan: Model based systems engineering and simulation for automotive systems development at GKN driveline, in: *NAFEMS World Congress 2017- A world of engineering simulation*, 11-14 June, Stockholm, Sweden (2017)

[Sch05] SCHENK, M.; STRAßBURGER, S. and KISSNER, H.: Combining virtual reality and assembly simulation for production planning and worker qualification, in: *Proc. of International Conference on Changeable, Agile, Reconfigurable and Virtual Production,* (2005)

[Sch09] SCHAMAI, W.; FRITZSON, P.; PAREDIS, C. and POP, A.: Towards unified system modeling and simulation with ModelicaML: modeling of executable behavior using graphical notations, in: *Proceedings of the 7th International Modelica Conference*, Como, Italy, (No. 043, pp. 612-621). Linköping University Electronic Press, December, (2009)

[Set11] SETH, A.; VANCE, J. M. andOLIVER, J. H.: Virtual reality for assembly methods prototyping: a review, in: *Virtual reality*, 15(1), 5-20, Sringer-Verlag London Limited, DOI 10.1007/s10055-009-0153-y, (2011)

[Sha19] SHAMSUZZOHA, A.; TOSHEV, R.; VU TUAN, V.; KANKAANPAA, T. and HELO, P. Digital factory–virtual reality environments for industrial training and maintenance, in: *Interactive Learning Environments*, 1-24 (2019)

[She02] SHERMAN, R. W. and CRAIG B. A.: Understanding Virtual Reality: Interface, Application, and Design. Morgan Kaufmann publishers 2002. *Chapter 1, pages 6-13*. ISBN: 1-55860-353-0 (2002)

[Sie16] SIEGEL, A., WEBER, C., MAHBOOB, A., ALBERS, A., LANDES, D. and BEHRENDT, M.: Virtual Acoustic Model for the Simulation of Passing Vehicle Noise, in: *ASME 2016 International Design Engineering Technical Conferences and Computers and Information in Engineering Conference* (pp. V01BT02A055- V01BT02A055), (2016)

[Sie17a] SIEGEL, Antje; WEBER, Christian; ALBERS, Albert; LANDES, David; BEHRENDT, Matthias: Simulation of Acoustic Product Properties in Virtual Environments Based on Artificial Neural Networks (ANN), in: *ICED – International Conference on Engineering Design 2017*, Vancouver, Canada (2017)

[Sie17b] SIEGEL, Antje; WEBER, Christian; ALBERS, Albert; LANDES, David; BEHRENDT, Matthias: Application of artificial neural networks for editing measured acoustical data for simulation in virtual environments, in: *59th Ilmenau Scientific Colloquium, Ilmenau*, 11-15 September, (2017)

[Sil11] SILHAVY, R.; SILHAVY, P. and PROKOPOVA, Z.: Behavioral modeling in system engineering, in: *13thWSEAS international conference on automatic control, modelling & simulation*, pp. 100-105, (2010)

[Sin11] SINDICO, A.; DI NATALE, M.; PANCI, G., Integrating SysML with Simulink using Open-source Model Transformations, in: *SIMULTECH* (pp. 45-56), July, (2011)

[Son18] SONG, H., CHEN, F., PENG, Q., ZHANG, J. and GU, P. (2018). Improvement of user experience using virtual reality in open-architecture product design, in: *Proceedings of the Institution of Mechanical Engineers, Part B: Journal of Engineering Manufacture*, 232(13), 2264-2275 (2018)

[Sta09a] STARK, R.; KRAUSE, F. L.; KIND, C.; ROTHENBURG, U.; MÜLLER, P. and STÖCKERT, H.: Competing in Engineering Design–the Role of Virtual Product Creation, in: *Proceedings of the 19th CIRP Design Conference–Competitive Design*. Cranfield University Press (2009)

[Sta09b] STARK, R.; HAYKA, H. and LANGENBERG, D.: New potentials for virtual product creation by utilizing grid technology, in: *CIRP annals – Manufacturing Technology, 58(1)*, 143-146 (2009)

[Ste04] STECKLEIN, J. M., DABNEY, J., DICK, B., HASKINS, B., LOVELL, R. and MORONEY, G.: Error cost escalation through the project life cycle. *NASA Johnson Space Center*. Url: https://ntrs.nasa.gov/archive/nasa/casi.ntrs.nasa.gov/20100036670.pdf [last accessed 30.03.2020] (2004)

[Ter15] TERKAJ, W., TOLIO, T. and URGO, M,: A virtual factory approach for in situ simulation to support production and maintenance planning, in: *CIRP Annals,* 64(1), 451-454 (2015)

[Vdi04] VDI-Guideline 2206: Design methodology for mechatronics systems, Berlin, Germany, (2004)

[Vdi86] VDI 2221:1986-11: Methodik zum Entwickeln und Konstruieren technischer Systeme und Produkte (1986)

[Vdi87] VDI-Guideline 2221: Systematic Approach to the Design of Technical Systems and Products", VDI, Düsseldorf, (1987)

[Vdi93] VDI 2221: Methodik zum Entwickeln und Konstruierentechnischer Systeme und Produkte (Systematic approach to the developmentand design of technical systems and products). Berlin:Beuth Verlag (1993)

[Vdi94] VDI/VDE 2422 : Entwicklungsmethodik für Geräte mitSteuerung durch Mikroelektronik (Systematical development ofdevices controlled by microelectronics). Berlin: Beuth Verlag (1994)

[Ver17] VERGARA, D., RUBIO, M. and LORENZO, M.: On the design of virtual reality learning environments in engineering, in: *Multimodal technologies and interaction*, 1(2), 11, (2017)

[Vre20] Virtual Robot Experimentation Platform (V-REP) by Coppelia Robotics – new name "CoppeliaSim". Url: http://www.coppeliarobotics.com/ [last accessed 30.03.2020]

[Vrm20] VRML – Virtual Reality Modelling Language, W3C. Url: https://www.w3.org/MarkUp/VRML/ [last accessed 30.03.2020]

[W3c20] W3C - State Chart XML (SCXML): State Machine Notation for Control Abstraction. https://www.w3.org/TR/scxml/ [last accessed 30.03.2020]

[Wal06] WALCZAK, K.: Beh-VR: modeling behavior of dynamic virtual reality contents, in: *International Conference on Virtual Systems and Multimedia* (pp. 40-51). Springer, Berlin, Heidelberg (2006)

[Web05] WEBER, Chrisitan: CPM/PDD – An Extended Theoretical Approach to Modelling Products and Product Development Processes, in: *Proceedings of the 2nd German-Israeli Symposium on Advances in Methods and Systems for Development of Products and Processes*, TU Berlin 07.-08.07.2005 pp. 159-179. Fraunhofer-IRB-Verlag, Stuttgart (2005)

[Web07] WEBER, Chrisitan: Looking at "DFX" and "Product Maturity" from the Perspective of a New Approach to Modelling Product and Product Development Processes, in: *Proceedings of the 17th CIRP Design Conference in co-operation with Berliner Kreis, "The Future of Product Development" (ed. by Krause, F.-L.), TU Berlin / Fraunhofer-Institut für Produktionsanlagen und Konstruktionstechnik(IPK),* 26.-28.03.2007, p. 85-104. Springer, Berlin-Heidelberg 2007.

[Web11] WEBER, Christian and HUSUNG, Stephan: Virtualisation of Product Development/Design - Seen from Design Theory and Methodology, in: *Culley, S.J.; Hicks, B.J.; McAloone, T.C.; Howard, T.J. & Reich, Y. (eds.), Proceedings of ICED 12, Vol. 2 (DS 68-2)*, pp. 226-235, the Design Society, Glasgow, (2011)

[Web16] WEBER, Christian; KRÖMKER, Heidi; Husung, Stephan; Hörold, Stephan; Mahboob, Atif and LIEBAL, Andreas: Benutzer- und aufgabenorientiertes virtuelles Modell für die Produktentwicklung, in: *EEE2016 – Konferenz Entwerfen Entwickeln Erleben 2016*, Dresden (2016)

[Wei15] WEILKIENS, Tim: SYSMOD - The Systems Modeling Toolbox, MBSE4U - Tim Weilkiens, (2015)

[Weg19] WiGeP: Wissenschaftliche Gesellschaft für Produktentwicklung, Ausgabe 1/2019, pp14

[Whi04] WHITMAN, L. E., JORGENSEN, M., HATHIYARI, K. and MALZAHN, D.: Virtual reality: its usefulness for ergonomic analysis, in: *Proceedings of the 36th conference on Winter simulation* (pp. 1740-1745). Winter Simulation Conference, December (2014)

[Ye07] YE, J., BADIYANI, S., RAJA, V. and SCHLEGEL, T.: Applications of virtual reality in product design evaluation, in: *International Conference on Human-Computer Interaction* (pp. 1190-1199). Springer, Berlin, Heidelberg, (2007)

[Zor03] ZORRIASSATINE, F.; WYKES, C.; PARKIN, R. and GINDY, N.: A survey of virtual prototyping techniques for mechanical product development, in: *Proceedings of the institution of mechanical engineers, Part B: Journal of engineering manufacture*, 217(4), 513-530, (2003)

Abbreviations and terms

Abbreviation	*Denomination*
ACT	Activity Diagram SysML
Actor	Actors are the persons with specific roles over the product life cycle e.g. fabricator, assembler or the end-user
API	Application programming interface
AR	Augmented Reality
BDD	Block Definition Diagram SysML
C#	C-Sharp object-oriented programming language
CAVE	Cave Automatic Virtual Environment
Context	A specific combination of environment and actor(s) in a specific product life-phase
CPM	Characterises Properties Modelling
CSM	Cameo Systems Modeler
DoF	Degrees of Freedom
dpi	"Dots per inch" can be understood as the number of pixels per inch
Environment	The surroundings of a product in a life-phase
FAS4M	Functional Architecture of Systems for Mechanical Engineers
FASP	Flexible Audio-visual Stereoscopic Projection system
fps	Frames per second
GUI	Graphical user interface
HLSA	High Level Solution Architecture
HMD	Head Mounted Display
IBD	Internal Block Definition Diagram SysML
MBSE	Model Based Systems Engineering
mecPro2	Modellbasierter Entwicklungsprozess cybertronischer Produkte und Produktionssysteme
ms	Milliseconds
OMG	Object Management Group
OOSEM	Object-Oriented Systems Engineering Method
PDD	Properties Driven Development
PLM	Product Lifecycle Management
PSS	Product Service System

Python control plugin	Python-based script to achieve bidirectional communication between SysML and V-REP
SE	Systems Engineering
SM	State Machine Diagram SysML
sVR	Smartphone Virtual Reality
SysML	Systems Modeling Language
SYSMOD	The Systems Modeling Toolbox
SysLM	System Lifecycle Management
TDM	Team Data Management
UDP	User Datagram Protocol
UML	Unified Modeling Language
Use case	Specific demand and interaction of environment and actor(s) with the product
VP	Virtual prototyping
VPD	Virtual product development
VRML	The Virtual Reality Modeling Language
VR-model	A model containing geometrical, structural and behavioural descriptions of a VR application consisting of product, actor and environment models
VR-scene	A scene in VR-software containing the geometrical objects along with their meta-data
VR-software	VR object modelling and rendering software
VR-system	CAVE, Powerwall, HMD, sVR
VR-user	The user of the VR-system (for the model developed in this thesis *"Product Designer"* is the VR-use)
V-REP	Virtual robot experimentation platform (Robot simulator). The new name "CoppeliaSim"
WFS	Wave Field Synthesis

List of publications with partial results

2019

MAHBOOB, Atif; HUSUNG, Stephan; WEBER, Christian; LIEBAL, Andreas and KRÖMKER, Heidi: Smartphone as a Stand-Alone Device for Rendering, Visualization and Tracking for Use During Product Development in Virtual Reality (VR), in: *International Design Engineering Technical Conferences and Computers and Information in Engineering Conference* (Vol. 59179, p. V001T02A079), *ASME* (2019)

MAHBOOB, A., HUSUNG, S., WEBER, C., LIEBAL, A., & KRÖMKER, H.: The Reuse of SysML Behaviour Models for Creating Product Use Cases in Virtual Reality, in: *Proceedings of the Design Society: International Conference on Engineering Design*, 1(1), 2021-2030. DOI:10.1017/dsi.2019.208 (2019)

MAHBOOB, Atif; HUSUNG, Stephan; WEBER, Christian; LIEBAL, Andreas and KRÖMKER, Heidi: Modellbasierter Systems Engineering Ansatz zur effizienten Aufbereitung von VR-Szenen, in: *EEE2019 – Konferenz Entwerfen Entwickeln Erleben in Produktentwicklung und Design 2019 Band 2*, Dresden, (2019)

LIEBAL, Andreas; KRÖMKER, Heidi; MAHBOOB, Atif; WEBER, Christian: Toolbox For User-Centered Specification Of VR Systems, in: *International Design Engineering Technical Conferences and Computers and Information in Engineering Conference* (Vol. 59179, p. V001T02A053), *ASME* (2019)

2018

MAHBOOB, A.; HUSUNG, S.; WEBER, C.; LIEBAL, A. and KRÖMKER, H.: An Approach for Building Product Use-Case Scenarios in Different Virtual Reality Systems, in: *ASME 2018 International Design Engineering Technical Conferences and Computers and Information in Engineering Conference* (pp. V01BT02A047-V01BT02A047). American Society of Mechanical Engineers (2018)

MAHBOOB, Atif; HUSUNG, Stephan; WEBER, Christian; LIEBAL, Andreas and KRÖMKER, Heidi: SYSML behaviour models for description of Virtual Reality environments for early evaluation of a product, in: *International Design Conference – Design 2018*, Faculty of Mechanical Engineering and Naval Architecture, University of Zagreb, (2018)

2017

MAHBOOB, A.; WEBER, C.; HUSUNG, S.; LIEBAL, A. and KRÖMKER, H.: Model based systems engineering (MBSE) approach for configurable product use-case scenarios in virtual environments, in: *DS 87-3 Proceedings of the 21st International Conference on Engineering Design (ICED 17) Vol 3: Product, Services and Systems Design, Vancouver, Canada, 21-25.08. 2017* (pp. 281-290) (2017)

MAHBOOB, A.; LIEBAL, A.; HUSUNG, S.; WEBER, C. and KRÖMKER, H.: A method for efficient and task oriented configuration of virtual reality (VR) models for the analysis of technical systems, in: *Proceedings of the 59th Ilmenau Scientific Colloquium (IWK), Ilmenau, Germany, September 11-15*. http://nbn-resolving.de/urn:nbn:de:gbv:ilm1-2017iwk-022:8 (2017)

LIEBAL, Andreas; MAHBOOB, Atif; WEBER, Christian; KRÖMKER, Heidi: CPM/PDD-basierter Ansatz für Produktevaluation in Virtual Reality (VR), in: *13. Magdeburger Maschinenbau-Tage 2017: autonom - vernetzt - nachhaltig*, Tagungsband. - Magdeburg: Universitätsbibliothek, (2017), S. 352-360. http://nbn-resolving.de/urn:nbn:de:gbv:ma9:1-10182 (2017)

2016

SIEGEL, A., WEBER, C., MAHBOOB, A., ALBERS, A., LANDES, D. and BEHRENDT, M.: Virtual Acoustic Model for the Simulation of Passing Vehicle Noise, in *ASME 2016 International Design Engineering Technical Conferences and Computers and Information in Engineering Conference* (pp. V01BT02A055-V01BT02A055). American Society of Mechanical Engineers (2016)

KRÖMKER, Heidi; WEBER, Christian; LIEBAL, Andreas; MAHBOOB, Atif; HÖROLD, Stephan and HUSUNG, Stephan: Die realen Seiten der virtuellen Produktentwicklung – Rollen- und aufgabenzentriertes Modell für Entwicklung von 3-D-Medienapplikationen, in: *Innteract Conference* (2016)

WEBER, C., HUSUNG, S., MAHBOOB, A., KRÖMKER, H., HÖROLD, S. and LIEBAL, A.: Benutzer-und aufgabenorientiertes virtuelles Modell für die Produktentwicklung. EEE2016 – Konferenz Entwerfen Entwickeln Erleben (2016)

MAHBOOB, Atif; LIEBAL, Andreas; HUSUNG, Stephan; WEBER, Christian and KRÖMKER, Heidi: Konzept für ein kontextabhängiges VR-Modell und die

Verhaltensbeschreibung in SysML am Produktbeispiel, in: *Virtuelle und Erweiterte Realität – 13. Workshop der GI-Fachgruppe VR/AR*, Bielefeld, Aachen: Shaker Verlag, ISBN 978-3-8440-4718-9, (2016), S. 149-152 (2016)

Annexure A (Codes)

***:red_udp* from figure 4-21**

```
import socket
import math
IP = "127.0.0.1"
Port = 5007
server_address = (IP, Port)
sock = socket.socket(socket.AF_INET, socket.SOCK_DGRAM)
sock.setsockopt(socket.SOL_SOCKET, socket.SO_REUSEADDR, 1)
sock.bind(server_address)
try:
  data, address = sock.recvfrom(1024) # buffer size is 1024 bytes
finally:
  sock.close()
array = data.split(" ")
joy_x = array[4]
joy_x = joy_x.split("]")
button = int(joy_x[0])
joy_x = joy_x[1]
joy_y = array[5]
joy_z = array[6]
y = joy_y
```

***:VR_Send from* figure 4-17**

```
import socket
import math
BUFFER_SIZE = 20
UDP_IP = "127.0.0.1"
UDP_Port = 8051
def udp_send(data):
  sock =socket.socket(socket.AF_INET, socket.SOCK_DGRAM)
  sock.sendto(data, (UDP_IP, UDP_Port))
def tcp(data):
  try:
     s = socket.socket(socket.AF_INET, socket.SOCK_STREAM)
     s.connect(('127.0.0.1',5555))
     s.send(data)
     print data
  finally:
     s.close()
ori_cleaner = ori_cleaner -90
ori_handle = -ori_handle
ori_handle = ("%.3f" % ori_handle)
ori_cleaner = ("%.3f" % ori_cleaner)
data_1 = '<trans | obj=cleaner | x='+ str((-pos_0*2.5))+ '| y=0.1 | z='+ str((-pos_1*2.5))
+'>'
data_2 = '<rot | obj=cleaner | x=0 | y='+str(ori_cleaner)+' |z=0>'
```

```
data_3 = '<rot | obj=handle | x='+str(ori_handle)+' | y=0 | z=0>'
data = ' '+str(data_1) +' '+ str(data_2) +' '+str(data_3)
MESSAGE = str(data)
tcp(data)
```

***:rec_pos_vrep* from figure 4-17**

```
import socket
import math
IP = "127.0.0.1"
Port = 23
server_address = (IP,Port)
sock = socket.socket(socket.AF_INET, socket.SOCK_DGRAM)
sock.setsockopt(socket.SOL_SOCKET, socket.SO_REUSEADDR, 1)
sock.bind(server_address)
try:
  data, address = sock.recvfrom(1024) # buffer size is 1024 bytes
finally:
  sock.close()
array = data.split(",")
pos_0 = array[0]
pos_0 = pos_0.split("[")
pos_0 = pos_0[1]
pos_1 = array[1]
ori_cleaner = array[2]
ori_handle = array[3].split("]")
ori_handle = ori_handle[0]
```

***:Rec_Simu* from figure 4-19**

```
import socket
import math
IP = "127.0.0.1"
Port = 22
server_address = (IP,Port)
sock = socket.socket(socket.AF_INET, socket.SOCK_DGRAM)
sock.setsockopt(socket.SOL_SOCKET, socket.SO_REUSEADDR, 1)
sock.bind(server_address)
try:
  data, address = sock.recvfrom(1024) # buffer size is 1024 bytes
finally:
  sock.close()
array = data.split(",")
Beuecherregal = array[0]
Beuecherregal = Beuecherregal.split("[")
Beuecherregal = int(Beuecherregal[1])
Bigsofa = int(array[1])
Bodenleiste_01 = int(array[2])
Bodenleiste_02 = int(array[3])
Bodenleiste_03 = int(array[4])
```

```
Cube0 = int(array[5])
Phonoschrank = int(array[6])
Plant = int(array[7])
Quader02 = int(array[8])
Quader04 = int(array[9])
Quader05 = int(array[10])
Quader69 = int(array[11])
Sessel = int(array[12])
Sessel__ottomane = int(array[13])
Stehlampe = int(array[14])
Tischplatt = int(array[15])
Tuer = array[16]
Tuer = Tuer.split("]")
Tuer= int(Tuer[0])
if Beuecherregal==0 and Bigsofa==0 and Bodenleiste_01==0 and Bodenleiste_02==0
and Bodenleiste_03==0 and Cube0==0 and Phonoschrank==0 and Plant==0 and
Quader02==0 and Quader04==0 and Quader69==0 and Sessel==0 and Sessel__otto-
mane==0 and Stehlampe==0 and Tischplatt==0 and Tuer==0:
  noone = 1
else:
  noone =0
```

Annexure B (Survey questionnaire)

Questionnaire about Smartphone VR *(Translated from the German language)*

Kindly provide us with answers of a few questions once you have tested both sVR applications. This questionnaire and the results will be completely handled anonymously.

1. Please choose your gender: (Please mark the relevant answer)

□ Male

□ Female

2. Which age group describes your age?

□ <25

□ 25-34

□ 35-44

□ 45-54

□ >54

2. Do you have past experience with VR and/or AR?

□ no

□ yes, and with

□ VR-HMDs (HTC Vive, Oculus Rift, PlayStation VR, …)

□ AR-HMDs (Microsoft HoloLens, …)

□ Powerwall, CAVE (from one to multiple projections walls)

□ Mobile-VR (360°-videos… on a tablet or a smartphone)

□ Mobile-AR (Extended catalogues, games… on a tablet or a smartphone)

□ others

3. What expectations do you have from Smartphone VR technology?

____

____

____

4. What expectations do you have from a smartphone VR application?

____

____

____

5. How do you rate the quality of the positional tracking based on the presented application in sVR?

very imprecise ○ ○ ○ ○ ○ very precise

6. Which application areas you consider most suitable for the application of sVR-system?

1. ____
2. ____
3. ____

7. Please take any two application area and individually evaluate the suitability of the display quality of the shown sVR-applications for your suggested areas of application.

Application area 1: ____

very poor ○ ○ ○ ○ ○ very good

Application area 2: ____

very poor ○ ○ ○ ○ ○ very good

8. Do you have any further comments/suggestions about the sVR-system?

____

____

Annexure C (SysML notations)

Table 15: SysML elements' notation and use

Symbol	Name	Explanation
	Block	To represent a system or sub-system with own structural and behavioural representations
	Interface Block	Defines the specification of interfaces e.g. flows, signals, flow directions etc.
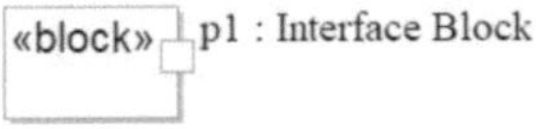	Standard port	Standard port typed by an interface block for communication with other elements
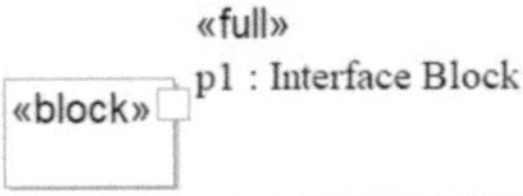	Full port	Full port typed by an interface block for communication with other elements
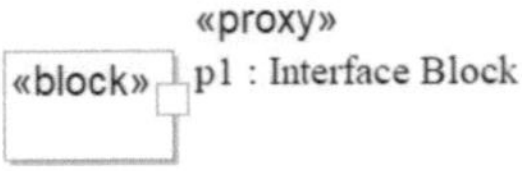	Proxy port	Proxy port typed by an interface block for communication with other elements
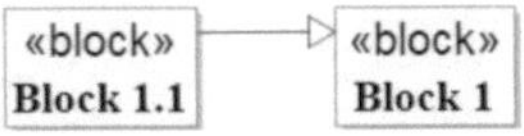	Generalisation relationship	Block 1 is a more generic version of Block 1.1. Block 1.1 inherits the interface specifications and ports from Block 1
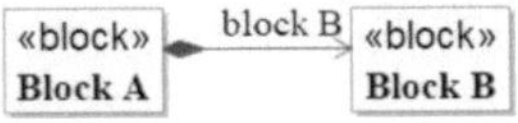	Directed composition	Block B is a component of Block A
	A Package	A package is used to organise model elements and diagrams
	Initial node	The start point of an activity diagram
	Activity Final	The endpoint of an activity diagram
	Control flow	Represents the flow of an execution token in an activity diagram

	Object flow	Represents the flow of object e.g. values in an activity diagram
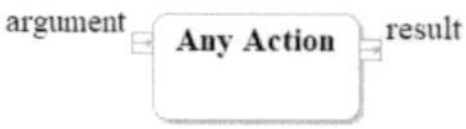	An Action	An Action with input "*argument*" and output "*result*" pins
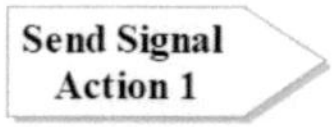	Send Signal Action	Used to generate a signal event
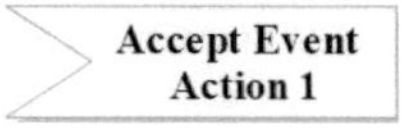	Accept Event Action	Used to simulate the reception of a signal or change event
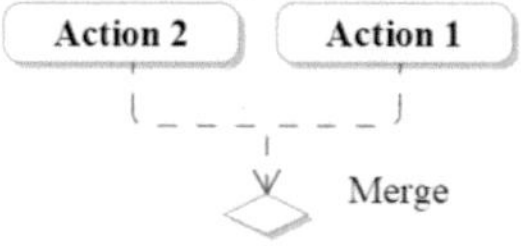	Merge	Used to merge multiple control flow into one
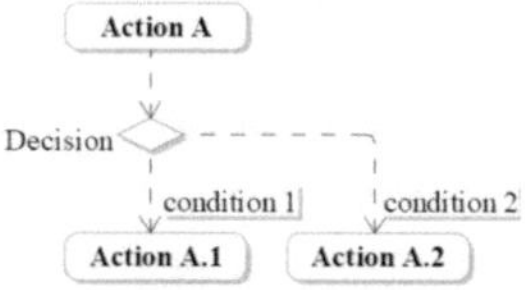	Decision	Used to split control flow against the fulfilment of some conditions
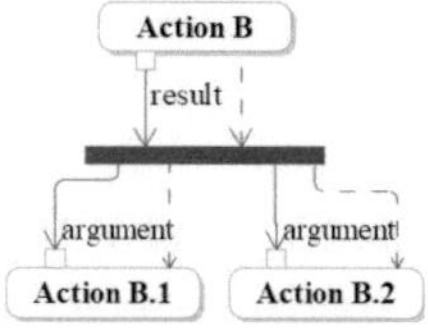	Fork horizontal	Used to split control and object flow into multiple flows
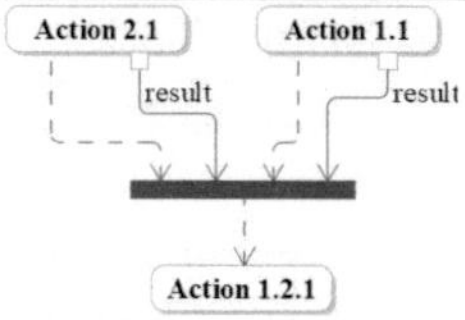	Join horizontal	Used to join multiple control or object flows
	Nested behaviour	Sign to indicate nested behaviour i.e. a diagram or behaviour element contains another behaviour element
	A state	A state with a transition arrow
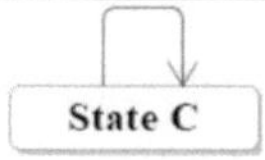	State with self-transition	A State that transits to its ownself after one execution

	Opaque action	Allows integration of object-oriented script directly inside the modelling (See also Annexure A)

Annexure D (Evaluation)

Table 16: Participants' profile

No.	Sector	Role	Age	VR experience	Study
1	Consumer goods manufacturer	Product developer	45-54	CAVE, HMD	A
2	Consumer goods manufacturer	Production manager	35-44	CAVE	A
3	Consumer goods manufacturer	Product developer	35-44	CAVE	A
4	Academia (VR-developer)	VR-developer/ programmer	35-44	CAVE, HMD, Smartphone	A
5	Academia (VR-developer)	Programmer	45-54	CAVE, HMD, Smartphone	A
6	Academia (Research)	Developer/ Programmer	35-44	CAVE	A
7	Academia (Research)	Researcher/ Product developer	35-44	CAVE, HMD	A
8	Machinery manufacturer	Product developer	24-34	Smartphone	A
9	Machinery manufacturer	Product developer	24-34	Smartphone	A
10	Academia (Research)	Researcher/ Product developer	25-34	CAVE, HMD, Smartphone	B
11	Academia (Research)	Researcher/ Product developer	25-34	HMD	B
12	Academia (Research)	Researcher/ Product developer	25-34	HMD	B
13	VR/AR Development Company	Developer/ Programmer	<25	HMD, Smartphone	B
14	VR/AR application development company	Developer / Programmer	<25	HMD, Smartphone	B
15	Academia (Research)	Developer/ Programmer	25-34	CAVE, HMD	B
16	Academia (Research)	Researcher/ Product developer	35-44	CAVE, Smartphone	B

17	Academia (Research)	Requirements engineer	35-44	HMD, CAVE, Smartphone, Powerwall	B

Table 17: Questionnaire A[93] *(Translated from the German language)*

No.	Question
A1	How useful is the virtual model for the evaluation of the coherence of product functionality?
A2	How useful is the virtual model for evaluating the coherence of product behaviour?
A3	Are all product relevant functions built in the shown application? This does not refer to the behaviour?
A4	Does the vacuum cleaner behave as expected?
A5	How realistic did you find the behaviour of the product?
A6	How realistic did you find carrying out the tasks with the product?
A7	How important is to build the realistic behaviour of the product?
A8	Are the product and environment model detailed enough for the evaluation of the conducted tasks?
A9	How well the product allows itself to be controlled in the desired way?
A10	How exact is the control of the product according to you?
A11	Would the indication of the collision of the product with its environment helpful in its current implementation?
A12	How important is the environment model for evaluating the coherence of product behaviour?
A13	How important is the environment model for the evaluation of the coherence of product functionality?
A14	How important is the environment model for evaluating the task practicality?
A15	How real did the environment appeal to?
A16	How useful is the virtual model for evaluating the task practicality?
A17	How did you find the speed of the VR-system?
A18	Did you have the feeling to be present inside the virtual world?

[93] This questionnaire was used inside an extended questionnaire (containing additional question about other evaluation aspects) for evaluation of a collective research project. To keep clarity, the complete extended questionnaire is not shown here and only the part relevant to this thesis is included.

A19	How close was your experience of the virtual environment to experiencing a real environment?
A20	How satisfied you were with the overall VR evaluation process?
A21	Could you experience any short-term mental or physical disabilities related to working in VR after the conducted tests?
A22	Would you prefer one of the technology over the other one for the evaluation of work goals or different work goals require different technologies [Possible answers: CAVE, HMD, it depends]?

Table 18: Questionnaire B *(Translated from the German language)*

No.	Question
B1	Are all the relevant functions of vacuum cleaner built?
B2	Does the vacuum cleaner behave as expected?
B3	How realistic did you find the behaviour of the product?
B4	How realistic did you find carrying out the tasks with the product?
B5	Are the product and environment model detailed enough for the evaluation of criteria and the conducted tasks?
B6	How well the product allows itself to be controlled in the desired way?
B7	How exact is the control of the product according to you?
B8	How did you find the speed of the VR-system?
B9	Did you have the feeling to be present inside the virtual world?
B10	How real did the environment appeal to?
B11	How close was your experience of the virtual environment to experiencing a real environment?
B12	How precise is the tracking according to you?
B13	How good is the display quality?
B14	How satisfied you were with the overall VR evaluation process?
B15	Could you experience any short-term mental or physical disabilities related to working in VR after the conducted tests?
B16	How do you rate the installation effort for the VR-system?
B17	Would you prefer one of the technology over the other one for the evaluation of work goals or different work goals require different technologies. Possible answers: o I will fundamentally prefer HMD o I will fundamentally prefer sVR o HMD is better suited for ____________________ o sVR is better suited for ____________________ o I would not use either of the technologies

General Questionnaire (Case Study A) *(Translated from the German language)*

Test person number.: Date:

A. Demographic information

1. Please specify your gender: (Please cross where suitable)
   - ☐ Male
   - ☐ Female

2. Please specify your age group: (Please cross where suitable)
   - ☐ <25
   - ☐ 25-34
   - ☐ 35-44
   - ☐ 45-54
   - ☐ >54

3. Please specify your professional position: (Please cross where suitable)
   - ☐ Product developer
   - ☐ Requirements Engineer
   - ☐ Constructor
   - ☐ Developer/Programmer
   - ☐ other, namely: ____________________

B. Virtual Reality

4. How much experience do you have in the use of Virtual Reality in a private capacity?

A lot of experience	++	+	-	--	No experience

4.1 If yes, which? (Please specify as detailed as possible)

5. How much professional experience do you have in dealing with Virtual Reality?

A lot of experience	++	+	-	--	No experience

5.1 If yes with which technologies?

☐ HMD (Head-Mounted-Displays e.g. Vive or Oculus Rift)
☐ Mobile VR (VR-glasses with Smartphone e.g. Samsung Gear)
☐ CAVE (multi-sided stereo projection)
☐ Powerwall (Stereo projection screen)
☐ other, namely: ______________________________

5.2 If yes, which models were part of the virtual environment?

☐ the product to be accessed
☐ the production related environment
☐ a product related digital human model

5.3 If yes, for what purpose? (Please specify as detailed as possible)

5.4 If yes, how? (Please specify as detailed as possible)

☐ for observation or assessment (passive)
  ☐ Observation through the eyes of a digital human model
  ☐ External observation
☐ for interaction (active)
☐ others, namely: (Please specify as detailed as possible)

6. How much experience do you have with 3D input devices (Flystick, 3D mouse, video game controller…)?

A lot of experience	++	+	-	--	No experience

7. What expectations do you have for working with virtual environments in the context of the desired task? (Please specify as detailed as possible)

General Questionnaire (Case Study B) *(Translated from the German language)*

Test person number: Date:

Start: End:

A. Demographic information

1. Please specify your gender: (Please cross where suitable)
   - ☐ Male
   - ☐ Female

2. Please specify your age group: (Please cross where suitable)
   - ☐ <25
   - ☐ 25-34
   - ☐ 35-44
   - ☐ 45-54
   - ☐ >54

3. Please specify your professional position: (Please cross where suitable)
   - ☐ Product developer
   - ☐ Requirements Engineer
   - ☐ Constructor
   - ☐ Developer/Programmer
   - ☐ other, namely: ____________________

B. Virtual Reality

4. How much experience do you have in the use of Virtual Reality in a private capacity?

A lot of experience	++	+	-	--	No experience

4.1 If yes, which? (Please specify as detailed as possible)

5. How much professional experience do you have in dealing with Virtual Reality?

A lot of experience	++	+	-	--	No experience

5.1 If yes with which technologies?

☐ HMD (Head-Mounted-Displays e.g. Vive or Oculus Rift)
☐ Mobile VR (VR-glasses with Smartphone e.g. Samsung Gear)
☐ CAVE (multi-sided stereo projection)
☐ Powerwall (Stereo projection screen)
☐ other, namely: ______________________________

6. How much experience do you have with 3D input devices (Flystick, 3D mouse, video game controller…)?

A lot of experience	++	+	-	--	No experience

7. What expectations/requirements do you have for working with VR technologies? (Please specify as detailed as possible)

Liste der bisher erschienenen Bände, Stand 02.12.2020

Bericht aus dem Institut für Maschinenelemente und Konstruktion (IMK), 1990 – 2010

Band 1 Institut für Maschinenelemente und Konstruktion der TU Ilmenau (Hrsg.):
Forschung und Lehre im Institut für Maschinenelemente und Konstruktion
(Institutsbericht)
Ilmenau : ISLE, 1999. - ISBN 3-932633-37-7

Band 2 Spiller, Frank:
Möglichkeiten der rechentechnischen Umsetzung von Erkenntnissen aus der Konstruktions-systematik unter Nutzung der Featuretechnologie
(Dissertation TU Ilmenau 1998)
Ilmenau : ISLE, 1998. - ISBN 3-932633-20-2

Band 3 Leibl, Peter:
Entwicklung eines featureorientierten Hilfsmittels für die Konstruktion kostengünstiger Produkte
(Dissertation TU Ilmenau 1998)
Ilmenau : ISLE, 1998. - ISBN 3-00-003695-4

Band 4 Lutz, Steffen:
Kennlinie und Eigenfrequenzen von Schraubenfedern
(Dissertation TU Ilmenau 2000)
Ilmenau : ISLE, 2000. - ISBN 3-932633-47-4

Band 5 Kletzin, Ulf:
Finite-Elemente-basiertes Entwurfssystem für Federn und Federanforderungen
(Dissertation TU Ilmenau 2000)
Ilmenau : ISLE, 2000. - ISBN 3-932633-48-2

Band 6 Volz, Andreas K.:
Systemorientierter Karosserie-Konzeptentwurf am Beispiel der Crashsimulation
(Dissertation TU Ilmenau 1998)
Ilmenau : ISLE, 2000. - ISBN 3-932633-52-0

Band 7 Brix, Torsten:
Feature- und constraint-basierter Entwurf technischer Prinzipe
(Dissertation TU Ilmenau 2001)
Ilmenau : ISLE, 2001. - ISBN 3-932633-67-9

Band 8 Rektor der TU Ilmenau und Institut für Maschinenelemente und Konstruktion der TU Ilmenau (Hrsg.) in Zusammenarbeit mit Carl Zeiss Jena GmbH
Vom Arbeitsblatt zum virtuellen Prototyp – 50 Jahre Konstruktionssystematik
(Institutsbericht)
Ilmenau : ISLE, 2002. - ISBN 3-932633-68-7

Band 9 Liebermann, Kersten:
Rechnergestütztes Entwurfs- und Optimierungssystem für Schraubendruckfedern

(Dissertation TU Ilmenau 2003)
Ilmenau : ISLE, 2003. - ISBN 3-932633-74-1

Band 10 — Meissner, Manfred; Denecke, Klaus:
Die Geschichte der Maschinenelemente als Fachgebiet und Institut an der Technischen Universität Ilmenau von 1953 bis 2003
(Institutsbericht)
Ilmenau : ISLE, 2003. - ISBN 3-932633-82-2

Band 11 — Geinitz, Veronika:
Genauigkeits- und auslastungsoptimierte Schraubendruckfedern
(Dissertation TU Ilmenau 2006)
Ilmenau : ISLE, 2006. - ISBN 3-938843-11-X

Band 12 — Institut für Maschinenelemente und Konstruktion (Hrsg.):
Festschrift zum Ehrenkolloquium anlässlich der Emeritierungen von Univ.-Prof. Dr.-Ing. habil. Dr. h.c. Günter Höhne und Univ.-Prof. Dr.-Ing. habil. Hans-Jürgen Schorcht
(Institutsbericht)
Ilmenau : ISLE, 2005. -ISBN 3-932633-97-0

Band 13 — Wittkopp, Tobias:
Mehrkörpersimulation von Schraubendruckfedern
(Dissertation TU Ilmenau 2005)
Ilmenau : ISLE, 2005. - ISBN 3-938843-07-1

Band 14 — Frank, Stefan:
Justierdrehen – eine Technologie für Hochleistungsoptik
(Dissertation TU Ilmenau 2007)
Ilmenau : ISLE, 2008. - ISBN 978-3-938843-35-4

Band 15 — Schilling, Thomas:
Augmented Reality in der Produktentstehung
(Dissertation TU Ilmenau 2008)
Ilmenau : ISLE, 2008. - ISBN 978-3-938843-42-0

Band 16 — Lotz, Markus:
Konstruktion von Messspiegeln hochgenauer Mess- und Positioniermaschinen
(Dissertation TU Ilmenau 2009)
Ilmenau : ISLE, 2009. - ISBN 978-3-938843-46-8

[Band 17] — Hackel, Tobias:
Grundlegende Untersuchungen zu vertikalen Positioniersystemen für Nanopräzisionsmaschinen
(Dissertation TU Ilmenau 2010)
Münster, Westf : Monsenstein & Vannerdat, 2010. -
ISBN 978-3-86991-111-3

[Band 18] — Frank, Thomas:
Konzeption und konstruktive Gestaltung der Messkreise von Nanomessmaschinen
(Dissertation TU Ilmenau 2010)
Münster, Westf : Monsenstein & Vannerdat, 2010. - ISBN 978-3-86991-194-6

Berichte aus dem Institut für Maschinen- und Gerätekonstruktion (IMGK), 2010 - ...

Band 19 Sondermann, Mario:
Mechanische Verbindungen zum Aufbau optischer Hochleistungssysteme
(Dissertation TU Ilmenau 2010)
Ilmenau : Univ.-Verl. Ilmenau, 2011. - ISBN 978-3-939473-94-7

Band 20 Husung, Stephan:
Simulation akustischer Produkteigenschaften unter Nutzung von Virtual Reality während der Produktentwicklung
(Dissertation TU Ilmenau 2011)
Ilmenau : Univ.-Verl. Ilmenau, 2012. - ISBN 978-3-86360-026-6

Band 21 Dobermann, Dirk:
Stabilisierung der Bildlage abbildender optischer Systeme
(Dissertation TU Ilmenau 2012)
Ilmenau : Univ.-Verl. Ilmenau, 2013. - ISBN 978-3-86360-056-3

Band 22 Taubmann, Peter:
Analyse der Ventilfederbewegung als Beitrag zur Beeinflussung der Verschleißursachen an den Auflageflächen
(Dissertation TU Ilmenau 2013)
Ilmenau : Univ.-Verl. Ilmenau, 2013. - ISBN 978-3-86360-059-4

Band 23 Erbe, Torsten:
Beitrag zur systematischen Aktor- und Aktorprinzipauswahl im Entwicklungsprozess
(Dissertation TU Ilmenau 2013)
Ilmenau : Univ.-Verl. Ilmenau, 2013. - ISBN 978-3-86360-060-0

Band 24: Ginani, Luciano Selva:
Optical Scanning Sensor System with Submicron Resolution
(Dissertation TU Ilmenau 2013)
Ilmenau : Univ.-Verl. Ilmenau, 2013. - ISBN 978-3-86360-068-6

Band 25: Heidler, Nils:
Untersuchungen zylindrischer Gasführungselemente für Hochvakuumanwendungen
(Dissertation TU Ilmenau 2015)
Ilmenau : Univ.-Verl. Ilmenau, 2016. - ISBN 978-3-86360-130-0

Band 26: Reich, René:
Möglichkeiten und Grenzen bei der Auslegung von Schraubendruckfedern auf Basis von Umlaufbiegeprüfungen
(Dissertation TU Ilmenau 2016)
Ilmenau : Univ.-Verl. Ilmenau, 2016. - ISBN 978-3-86360-139-3

Band 27: Resch, Jens:
Kontextorientierte Entwicklung und Absicherung von festen Verbindungen im Produktentstehungsprozess der Automobilindustrie
(Dissertation TU Ilmenau 2016)
Ilmenau : Univ.-Verl. Ilmenau, 2016. - ISBN 978-3-86360-143-0

Band 28: Scheibe, Hannes:
Aktiv-adaptive Polierwerkzeuge zur Herstellung rotationssymmetrischer Asphären

(Dissertation TU Ilmenau 2016)
Ilmenau : Univ.-Verl. Ilmenau, 2016. - ISBN 978-3-86360-147-8

Band 29: Reeßing, Michael:
Softwarewerkzeuge für den phasen- und domänenübergreifenden Entwurf
(Dissertation TU Ilmenau 2016)
Ilmenau : Univ.-Verl. Ilmenau, 2017. - ISBN 978-3-86360-169-0

Band 30: Lux, Rüdiger:
Wärmebehandlung von SiCr-legiertem, ölschlussvergütetem Federstahldraht
(Dissertation TU Ilmenau 2018)
Ilmenau : Univ.-Verl. Ilmenau, 2018. - ISBN 978-3-86360-185-0

Band 31: Thomisch, Marco:
Methodik zur Bestimmung optimaler Parameter beim Drahtumformen
(Dissertation TU Ilmenau 2018)
Ilmenau : Univ.-Verl. Ilmenau, 2018. - ISBN 978-3-86360-187-4

Band 32: Wohlfahrt, Fabian:
Kraftgesteuerte Messzelle für Dilatometeranwendungen
(Dissertation TU Ilmenau 2018)
Ilmenau : Univ.-Verl. Ilmenau, 2019. - ISBN 978-3-86360-193-5

Band 33: John, Kerstin:
Untersuchung von Umlenkelementen zur Anwendung in der interferometrischen Längenmesstechnik
(Dissertation TU Ilmenau 2018)
Ilmenau : Univ.-Verl. Ilmenau, 2019. - ISBN 978-3-86360-197-3

Band 34: Mack, Benjamin:
Untersuchungen zum Schadensmechanismus Torsionsschwingbruch durch Kontaktermüdung an Schraubendruckfedern
(Dissertation TU Ilmenau 2018)
Ilmenau : Univ.-Verl. Ilmenau, 2019. - ISBN 978-3-86360-198-0

Band 35: Hesse, Miriam:
Untersuchung der Absicherung von Montageprozessen am Beispiel von Produktionsanläufen in der Automobilindustrie
(Dissertation TU Ilmenau 2019)
Ilmenau : Univ.-Verl. Ilmenau, 2020. - ISBN 978-3-86360-221-5

Band 36: Scheler, Marcel:
Auswahl robuster Wirkprinzipien auf Basis einer Erweiterung des CPM/PDD-Ansatzes
(Dissertation TU Ilmenau 2020)
Ilmenau : Univ.-Verl. Ilmenau, 2020. - ISBN 978-3-86360-225-3

Band 37: Schienbein, Ralf:
Grundlegende Untersuchungen zum konstruktiven Aufbau von Fünfachs-Nanopositionier- und Nanomessmaschinen
(Dissertation TU Ilmenau 2020)
Ilmenau : Univ.-Verl. Ilmenau, 2020. - ISBN 978-3-86360-229-1

Band 38: Mahboob, Atif:
Modelling and use of SysML behaviour models for achieving dynamic use cases of technical products in different VR-systems
(Dissertation TU Ilmenau 2020)
Ilmenau : Univ.-Verl. Ilmenau, 2021. – ISBN 978-3-86360-234-5